Voices and Echoes for the Environment

POWER, CONFLICT, AND DEMOCRACY
American Politics into the Twenty-first Century
Robert Y. Shapiro, Editor

Power, Conflict, and Democracy:
American Politics Into the Twenty-first Century
Robert Y. Shapiro, Editor

This series focuses on how the will of the people and the public interest are promoted, encouraged, or thwarted. It aims to question not only the direction American politics will take as it enters the twenty-first century but also the direction American politics has already taken.

The series addresses the role of interest groups and social and political movements; openness in American politics; important developments in institutions such as the executive, legislative, and judicial branches at all levels of government as well as the bureaucracies thus created; the changing behavior of politicians and political parties; the role of public opinion; and the functioning of mass media. Because problems drive politics, the series also examines important policy issues in both domestic and foreign affairs.

The series welcomes all theoretical perspectives, methodologies, and types of evidence that answer important questions about trends in American politics.

Voices and Echoes for the Environment

Public Interest Representation in the 1990s and Beyond

RONALD G. SHAIKO

COLUMBIA UNIVERSITY PRESS

NEW YORK

Columbia University Press
Publishers Since 1893
New York   Chichester, West Sussex
Copyright © 1999 Columbia University Press
All rights reserved

Library of Congress Cataloging-in-Publication Data

Shaiko, Ronald G., 1959–
    Voices and echoes for the environment : public interest
representation in the 1990s and beyond / Ronald G. Shaiko.
        p.   cm. — (Power, conflict, and democracy)
    Includes bibliographical references.
    ISBN 0-231-11354-4 (cl). — ISBN 0-231-11355-2 (pa)
    1. Environmental policy—United States—Societies, etc.
2. Environmental policy—United States—Citizen participation.
3. Public interest—United States.   I. Title.   II. Series.
GE180.S42   1999
363.7'0525'0973—dc21   98-52088
                                                                 CIP

Casebound editions of Columbia University Press books are printed on permanent and
durable acid-free paper.
Printed in the United States of America
c 10 9 8 7 6 5 4 3 2 1
p 10 9 8 7 6 5 4 3 2 1

For My Parents

It's enough to be on your way.
It's enough just to cover ground.
It's enough to be moving on.

—JAMES TAYLOR, MEMBER OF THE BOARD OF TRUSTEES,
NATURAL RESOURCES DEFENSE COUNCIL

Contents

Today, public interest representation in the United States is a multibillion dollar political enterprise. Yet only twenty-five years ago, the public interest sector at the national level amounted to no more than one hundred organizations with combined budgets totaling less than one hundred million dollars. This study focuses on the consequences of the dramatic organizational transformation of a loosely organized collection of social movement groups in the 1960s into the highly organized, professionalized public interest organizations of the late 1990s with specific attention given to the organization, maintenance, and representation of environmental groups.

As memberships, budgets, and staffs rise and fall, organization leaders are faced with new, often competing goals. No longer can it be assumed that the primary goal of public interest organizations is effective policy influence. Organizational maintenance now encroaches on the goal orientation of many organizations. In the new age of high-technology mass marketing, including the use of the "information superhighway" and professionalized staffing, public interest representation has a new meaning. Organization leaders now must provide effective political representation in the policy-making process as well as maintain organization infrastructures. Most contemporary public interest organizations have great difficulty achieving both these goals. For social scientists and students interested in political representation, for public interest leaders attempting to address these issues, and for the millions of citizens in this country who contribute billions of dollars annually to the public interest sector, this is a serious problem.

The title of this book is derived from two sources: one is an eminent political scientist who wrote numerous books and articles that have become classics in the field, and the other is an interview subject in the research for this book. In his posthumously published book, *The Responsible Electorate*, V. O. Key Jr. conceptualized the relationship between political parties and their candidates and the American electorate in the following manner: "The voice of the people is but an echo. The output of an echo chamber bears an inevitable and invariable relationship to the input. As candidates and parties clamor for attention and vie for popular support, the people's verdict can be no more than a selective reflection from among the alternatives and outlooks presented to them."[1] A similar relationship exists between leaders of national public interest organizations and their members as well as potential members, both in the recruitment and retention of an attentive followership and in the mobilization of members by leaders for political purposes.

This latter aspect of the relationship became quite clear during an interview with Joel Thomas of the National Wildlife Federation (NWF). As we discussed the level of policy content in the communications between NWF leaders and their members, Thomas argued that in a large organization like NWF informing the membership on a variety of policy concerns causes an inevitable feedback from members that NWF leaders do not wish to facilitate: "If they [organization leaders] are sending out good communications, to some degree, they are going to end up talking to themselves. There is a real danger, if you have an undifferentiated mass of members out there, where you have a one-on-one relationship—you to them. Who are you talking to? Or are you just hearing the echo of your own voice?"

The membership of most public interest organizations, however, is far more homogeneous than that of NWF. As such, the ability of the leaders to inform their members so that they may actively pursue their organization's agenda from the grassroots is crucial. For organization leaders (*voices*) who believe that an informed membership (*echoes*) is a powerful political resource, it is important that both components work in concert. As public interest leaders and their lobbyists seek to influence the federal policy-making process directly in Washington, D.C., it has become increasingly vital that they also keep their membership informed about ongoing policy debates in order that members may echo the sentiments of the leadership from the grassroots.

In this book, through the analysis of organizational attributes, leadership styles, leadership communications with members, recruitment efforts, membership motivations, and leadership-membership political activities, using a variety of methodologies, the relationships between leadership (*voices*) and membership (*echoes*) are defined and assessed. Substantively, the study begins with a broad overview of the political context within which public interest organizations are created and maintained. Chapter 1 provides a broad comparative analysis of the transformation of the public interest sector from the 1960s through the 1990s using organizational data collected by Jeffrey Berry and the Foundation for Public Affairs. With the broad public interest organizational patterns established, the remaining chapters include analyses of one component of the public interest sector— national environmental organizations.

Chapter 2 begins with an overview of the development of the major national environmental organizations active in the federal policy-making process today as well as the changing political climate in public opinion toward environmentalism. This is followed by descriptive analyses of five national environmental organizations: Environmental Defense Fund, National Wildlife Federation, Sierra Club, The Wilderness Society, and one fledgling organization that recently closed its doors after twenty-six years of environmental advocacy, Environmental Action. The analyses presented in chapters 3, 4, 5, and 6 will focus on these five organizations.

The changing roles of organizational leadership will be analyzed in chapter 3. Throughout the environmental movement, group leaders have wrestled with the often competing tasks of organizational maintenance and political representation. Each of the leaders of these five organizations has attempted to deal with these issues, with varying degrees of success. Similar leadership challenges within other environmental organizations will be discussed as well, including the ability of these organizations to work in concert through coalitional activities. Leadership styles and management approaches are analyzed in order to assess the overall effectiveness of organizational maintenance.

The specific approaches to membership recruitment and retention, including direct mail and telemarketing, are analyzed in chapter 4. Additional fund-raising efforts, such as corporate, government, and foundation solicitations, are also analyzed. The recruitment and retention strategies of each of the five environmental organizations are assessed to determine the effectiveness of their organizational maintenance efforts.

Chapter 5 begins the exploration of the relationships between organizational leaders and their members. A multi-incentive model of membership motivation and mobilization is presented and analyzed using membership data from the five environmental organizations identified earlier. Across these five organizations, as well as in the larger movement, organizational leaders attempt to appeal to potentially interested citizens with diverse sets of incentives (e.g., the policy goals of the organization; an attractive magazine; outings programs; discounts on books, clothing, trips; local meetings with other members) to induce them to join their particular organizations. Beyond soliciting new members, leaders must be equally concerned with retaining current members. Using survey responses from members of the five organizations, the motivations of individual members for belonging to specific organizations will be explored, and linkages between leadership incentives and membership motivations will be identified.

Chapter 6 further explores the linkages between organizational leaders and members by analyzing the content of the communications between leaders and members and connecting the issue agendas of leaders with those of activist members. The messages sent from leaders to members are analyzed to discern whether or not coherent policy agendas are being presented to active members in order to mobilize them to support or oppose specific policy issues being addressed by the federal government. The increasing use of the information superhighway by environmental leaders and members is also analyzed.

Finally, in chapter 7, the broad organizational relationships between organizational maintenance and political representation are assessed for the national environmental movement in the United States at the end of the twentieth century. Specific conclusions based on this assessment are then applied to the larger public interest sector. Coming full circle, this study looks at the larger public interest sector and addresses the consequences of these findings for the larger collection of public interest organizations that seek to influence the federal policy-making process.

To discern the various relationships between organizational leaders and members, I began interviewing leaders of public interest organizations for this project in 1985–1986 in Washington, D.C. (as well as in San Francisco and New York City), while conducting my dissertation research at the Brookings Institution, and completed the final round of interviews a decade later. These leaders included executive directors, board members, membership/development directors, policy staffers, and lobbyists. The interviews lasted approximately one hour on average, ranging from forty minutes to more than three hours. Virtually all interviews were conducted in person with follow-up interviews made by telephone, particularly when I was not in Washington from 1987 to 1990. Since that time, the majority of follow-up interviews were conducted in person with organizational personnel in Washington. Without the cooperation of these organization leaders, this study would not have been possible.

In addition to interviewing the leadership of these organizations, several other methodologies were employed in order to address these issues from various perspectives. For each methodological approach, I received significant help in collecting and analyzing various types of data. In chapter 1, data on organizational attributes are compared. The first source of organizational data is Jeffrey Berry's *Lobbying for the People*. In 1972 Jeffrey Berry gathered data on eighty-three public interest organizations. His analysis of these organizations represents the first systematic research on the public interest sector. His sample provides an accurate measure of the state of the sector in the early 1970s. The other, more recent sources of public interest organizational data are derived from three volumes of *Public Interest Profiles*; each volume includes organizational information for 250 public interest organizations. These data sets were collected in 1985, 1990, and 1995 by the Foundation for Public Affairs. (See appendix A for listings of organizations included in each of the three data sets.) The latter data source, published in 1996, is the most recent compilation of organizational data by the Foundation for Public Affairs. Environmental organizational data presented in chapter 2 were derived from the archives of the Foundation for Public Affairs as well as from the environmental organizations themselves through interviews and the sharing of financial and organizational documents. The analyses in chapters 5 and 6 focus on the motivations of individual citizens for belonging to one or more of the five environmental organizations. The data for these analyses were collected by Robert Cameron Mitchell for Resources for the Future in 1978. While this membership survey research is somewhat dated, it remains one of the richest sources of data on membership motivations collected to date.

The final methodology utilized in this study is content analysis. The content of both the communications sent from leaders of environmental organizations to members (magazines, newsletters) and the direct mail used in recruiting new members was analyzed. In order to analyze, one needs content. The editors of the organization publications, in addition to being interview subjects, were also helpful sources for tracking down back issues of magazines and other communications. As for the content analysis of direct mail pieces, literally hundreds of organizations have made the research presented in chapter 4 possible through their generous outpourings of direct mail solicitations.

These various methodological approaches lead to the conclusion that organizational effectiveness in the public interest sector requires political leadership on the part of organizational elites. It requires a restructuring of public interest organizations that places policy influence ahead of organizational maintenance and also a reorganizing of public interest organizations in order to get the most out of limited human and financial resources. Above all, effective political influence in a variety of policy arenas, from local to global, requires that organization leaders serve their constituencies by "acting with" their members in a more concerted effort to provide public interest representation.

ACKNOWLEDGMENTS

It is difficult to know where to begin when attempting to summarize in a few hundred words my thoughts about this book. Perhaps it is best to follow the chronology of events that led to its publication by Columbia University Press. This project began as my dissertation for the political science department at the Maxwell School of Citizenship and Public Affairs at Syracuse University. While this book bears little resemblance to the dissertation defended almost a decade ago, the initial ideas and approaches laid out in that work served as building blocks for my subsequent research. That my dissertation work served me well in my future research is a credit to my core committee members, Linda Fowler, now at Dartmouth, Tom Patterson, now at Harvard, and Jeff Stonecash. From each of these professors, colleagues, and friends, I have learned much. I am grateful for their time and effort during the early stage of my career. Beyond these key faculty members, I also benefited from the support of the Maxwell School through its Maxwell Alumni Dissertation Fellowship. In addition, I would like to thank Paul Peterson, then Director of Governmental Studies at the Brookings Institution, for providing support for my dissertation research at Brookings for the 1985–1986 academic year.

During the past decade, I have been influenced by the scholarship and intellectual stimulation of a number of colleagues, past and present, including Lynette Rummell, Rhys Payne, Jon Hale, Joe Aistrup, Gregg Ivers, Diane Singerman, Christine DeGregorio, and John Boiney. Special thanks to David Rosenbloom, who has served as an important source of support, first as a professor and now as a colleague. In addition to these friends, there are several additional scholars, colleagues, and friends to whom I am intellectually indebted. Their work on interest groups has guided and inspired me; the imprint of each will be recognizable in the pages that follow. My thanks and gratitude go to Jeff Berry, Chris Bosso, Mark Hansen, Larry Rothenberg, Terry Moe, Andrew McFarland, Robert Mitchell, Burdett Loomis, Allan Cigler, Russ Dalton, and the late Jack Walker.

At various stages of completion, this manuscript was read by and benefited from the comments of Jeff Berry, Russ Dalton, Chris Bosso, and Holmes Rolston. Numerous others read and commented on parts of the manuscript when it was still in conference paper form. Once the manuscript reached Columbia University Press, the path to publication could not have been smoother. Thanks to Bob Shapiro for his initial interest in the manuscript for his series, "Power, Conflict, and Democracy: American Politics into the Twenty-First Century." John Michel, along with Alexander Thorp, shepherded the manuscript with great care and attention through the review and production processes. Steven Cohen of Columbia University and the additional anonymous reviewers provided prompt, constructive, and supportive comments that made the manuscript substantively stronger. Sabine Seiler performed the copyediting that made the manuscript more readable.

Along the way, a number of my graduate students also helped with various aspects of my research on this project. Whether serving as coders for the content analyses, fact-checkers, data managers, or as word-processing/computer gurus in generating tabular material, each of the following past or present graduate students made my life easier through their assistance. My thanks to Tim Huelskamp, Dale Weighill, John McCormick, Mark Hutchins, and Marc Wallace for their time and effort.

In addition to these academicians, there are two groups of individuals who deserve my recognition and thanks. First, without the cooperation of the more than 100 leaders and staffers of environmental and other public interest organizations this project would have been much more limited in scope. The variety of perspectives offered in interviews with these activists greatly informed my judgments regarding public interest representation in American politics today. Second, the environmental journalists throughout the country who analyze and write about environmental politics and policy on a regular basis have also informed my judgments. Evidence of their influence on my thinking is found in the chapter notes and in the bibliography.

Finally, I have dedicated this book to my parents, Tom and Joan Shaiko. Over the years, they have provided the stability, support, and love that have saved me from taking the pitfalls of academic life too seriously. Thanks, Mom and Dad.

Voices and Echoes for the Environment

Voices and Echoes in the Public Interest Marketplace

The Development of the Public Interest Sector

> Students of history forecast a surge of activism in the 1990s. They believe
> there are thirty-year cycles in American history for such surges—the '30s,
> the '60s, and now the '90s. There does seem to be a backlog of mounting
> social problems inviting attention beyond the routine and redundant.
> And the younger generations are becoming more civic-minded after a
> decade of self-centered grasps for gold. Perhaps more significant is the
> diversification of citizen strategies and the maturation of their efforts.
>
> —RALPH NADER

When consumer activist and public interest entrepreneur Ralph Nader first
arrived on the Washington political scene more than three decades ago, his
vision of public interest activism was distinctly social in nature.[1] He believed that people mattered and that their collective political activism could
change public policy outcomes. Public interest activism meant social networking and "taking it to the streets," literally. Today, public interest activism, while still incorporating citizen activism to some extent, often takes
the form of individual citizens, solicited by high-technology mass recruitment efforts, writing membership checks once a year to their favorite cause
groups in return for monthly newsletters or magazines.[2]

While citizen activists once may have taken to the streets to demonstrate their critical mass of support for or opposition to government policies, today this same generation, politically socialized in the 1960s, is able
to pay an organization or organizations to protest or get arrested for them.[3]
The success of Greenpeace, a multimillion dollar international environmental organization, serves as the best example of this phenomenon in the
1990s. Many baby boomers, now approaching fifty, have neither the time

nor the inclination to place themselves in harm's way as they might have done twenty or thirty years ago. Now they have responsibilities of family and work that preclude such direct activism. Nonetheless, many of these graying citizens do have sufficient disposable income to support such direct action efforts financially.

Having not totally jettisoned their earlier convictions, their option in the 1990s is to contribute to organizations that will "take it to the streets" for them. Greenpeace serves this function for hundreds of thousands of contemporary citizen activists. Leaders of Greenpeace understand this relationship quite well. Peter Bahouth, former executive director of Greenpeace, referred to this connection as "the vicarious thrill of living through the group's activists who do things that members themselves would not do."[4]

Today's teens and twenty-somethings, the current generation of potential citizen activists, whether labeled Generation X or Thirteenth Generation, have not demonstrated their political activism in the 1990s in the same way as many of their parents or Ralph Nader did a generation earlier. Social causes must be packaged differently for this generation, whether through concert events and CDs or through the internet. Organizational leaders in the public interest sector in the late 1990s must deal with these diverse groups of members and potential members—with the baby boomers and older people for their organizational livelihood today and with the next generation for the viability of public interest representation into the twenty-first century.

Today, public interest political activism, whether practiced by baby boomers and senior citizens or by those citizens in their teens, twenties, or thirties, has a tendency to be less social and spontaneous and is more likely to be orchestrated by organization leaders managing multimillion dollar political enterprises through the direct-mail mobilization of individual citizen actions such as letter-writing, faxing, e-mailing, or telephoning a public official—grassroots lobbying in the 1990s. This opening chapter explores the path from social movements to institutionalized, professionalized public interest organizations, a path traveled by public interest leaders and their affiliated member activists over the past twenty-five years. In no other political sector of society have there been such dramatic quantitative and qualitative changes during that time as in the public interest sector. Public interest organizations now occupy significant positions in the political marketplace and are more diverse in substance and ideological perspective than ever before in American history. According to Ralph Nader, "these groups are diverse and differ greatly in policy, effectiveness, strategy, and stamina. They only begin to cover the wide range of civic activity, organized and unorganized, that operates daily in our country. And they do not all align themselves on one end of the political, economic, or social philosophy spectrums."[5]

The public interest social movements of the 1960s have been transformed into contemporary public interest organizations and have been joined by scores of new political interests in the larger, more diverse political marketplace. At the same time these new organizations are competing with older, established public interest enterprises for activist supporters and their financial resources in an economic marketplace that provides the foundation for public interest activism.

In an interesting assessment of the public interest movement, David Cohen, former president of Common Cause and current codirector of the Advocacy Institute, revisits the doctrine of American exceptionalism, a revolutionary and frontier era philosophy that stressed the unique, unbridled capacity of Americans to succeed in all endeavors, and argues that it is quite applicable to contemporary public interest efforts:

> American exceptionalism is about the capacity for invention. The capacity of the public interest world to adopt rapid communication technology to inform its constituency begat its imitations in all other interest groups. A meeting of corporate officers on applying grassroots lobbying pressure now is no different functionally from that of a public interest lobby group. The subjects and stands differ; the techniques do not.[6]

Another manifestation of public interest inventiveness is the growing use of political coalitions in order to influence policy outcomes. In the 1990s these coalitions not only include allied public interest organizations but also business and/or labor interests, depending on the issues in response to which the coalitions are organized. These two techniques, grassroots lobbying and coalition building, are the most often identified strategies offered by organization leaders as influence methods for the 1990s. From environmental organizations to consumer groups and from economic organizations as diverse as the American Medical Association and the AFL-CIO to the National Association of Realtors and the National Association of Home Builders, grassroots efforts and coalitional strategies are now significant components of the interest group arsenal.[7] The significance of these methods is underscored by the unwillingness of Congress to address them in the Lobbying Disclosure Act of 1995, passed in the 104th Congress and signed by President Clinton. In fact, the lobby reform efforts in the 103rd Congress were killed, in part, because there were grassroots provisions included in the reform bill. The phrase "grassroots lobbying" does not appear in the lobbying law that went into effect on January 1, 1996, and many coalitional activities fall outside of the jurisdiction of the new law.[8]

While both of these strategies have their broadest application in the public interest sector, their increased use across the wide spectrum of organized interests prompts a closer look at these methods, particularly at how they are employed by public interest organizations. Both methods assume certain relationships within organizations as well as between organizations. In economic organizations, internal relationships are qualitatively different from those between leaders and members of public interest groups. In the former, an economic relationship binds workers and their employers together, union members and their leaders, or professionals and their associations. With this direct economic linkage, the relationship is necessarily different from the one that binds public interest leaders to their members or supporters. It is ironic that the broad array of economically driven organizations has lagged behind public interest groups in mobilizing its adherents, given the closer economic link (and, at the extreme, coercive relationship) between leaders and followers.

The analyses that follow seek to assess some of the assumptions made when public interest organizations attempt to employ both grassroots lobbying and coalition lobbying. Grassroots lobbying, presumably, is based on some degree of congruence between the positions staked out by organization leaders and the opinions held by their members. And if such a connection exists between leaders and members of public interest organizations, there must also exist similar levels of issue salience. Issue agendas in these groups are determined through many different processes. Some organizations involve members directly in the agenda-setting process, while others employ no such membership-linked input. For the groups relying on professional staff to construct policy agendas it may be a significant leap of faith to assume that their members hold similar opinions and, more important, are willing to act upon them. It is from these diverse relationships between leaders and members that the title of this book is derived. In order to have effective advocates in the policy process, the messages sent directly to policy makers from organization leaders and their lobbyists—the *voices*—must be supported by similarly informed messages from the grassroots memberships—the *echoes*.

Citizens who belong to public interest organizations are likely to be political activists, generally defined. According to David Cohen, "Voting alone does not satisfy many Americans as a sole basis of representation. So Americans, with their greater education and information, do something about it."[9] This sector of society contributes to campaigns, volunteers for political activities, writes letters to public officials, and belongs to public interest groups, thereby providing political legitimacy for such organizations as well as securing their financial stability through direct contributions.[10] Throughout this work, the attitudes, opinions, and activities of these citizen activists and their unique (political/financial) relationships with organizational leaders are analyzed. Similarly, the leadership structures, processes, and styles identified in public interest organizations are explored. Grassroots lobbying strategies assume a significant relationship between organizational leaderships (*voices*) and their corresponding memberships (*echoes*). Coalitional behavior assumes that there is sufficient coherence of policy management within public interest organizations as well as a communication process within the public interest community and beyond to allow for significant organizational linkage. These assumptions will be explored in the chapters that follow to provide a comprehensive assessment of the state of organizational effectiveness in the public interest sector.

From Social Movements to Political Organizations

Public interest representation has become a big business in the United States. During the past twenty-five years public interest movements have developed into major political enterprises. Today, the public interest sector includes more than 3,000 national nonprofit organizations representing a myriad of substantive and ideological positions.[11] "The growth in number and influence" of public interest organizations "is one of the most striking developments in interest group politics in the United States."[12] Beyond the mobilization of interests at the national level, tens of thousands of regional, state, and local organizations are now providing public interest representation in many new political arenas. Together, these organizations are supported through memberships and contributions of more than 80 million individuals, whose giving totals more than $4 billion annually. In addition to the support of individual citizens, private foundations and corporations, as well as governmental agencies, provide significant financial support for public interest advocacy.[13]

In less than three decades, the public interest movement has been transformed into a collection of autonomous, often competing organizations, each seeking to represent its own particular version of "the public interest."

While Schlozman and Tierney concluded from their analysis of interest group data collected more than fifteen years ago that the most accurate description of the changes in the entire interest group sector was "more of the same,"[14] the changes that have taken place in the public interest sector are not simply quantitative in nature. The transformation in the organization and maintenance of public interest organizations has resulted in a qualitative change in the nature of public interest representation.[15]

Several authors have chronicled the development of the public interest movement since the turn of the century. Scholars writing in the 1950s and 1960s believed that mobilizations of the citizenry were a response to "disturbances" in society, including the general increase in the complexity of working relationships and more specific imbalances caused by wars, fluctuations in business activity, and the increased role of government in everyday life.[16] Truman's "disturbance theory," which stood as conventional wisdom for twenty years, was later challenged by Salisbury's "exchange theory" of interest group development. Rather than stressing market disturbances as catalytic forces in the formation of groups, Salisbury focused more broadly on the origins and maintenance of organized interest groups. By doing so he was able to identify the role of the leader or political "entrepreneur" as an important factor in determining the success or failure of such groups.[17] Jeffrey Berry has offered a test of these two theories. He found that two-thirds of the public interest groups in his study "were begun by entrepreneurs working without significant disturbances as additional stimuli."[18] In the mid-1970s, Andrew McFarland offered a "civic balance theory" of public interest representation that incorporates some aspects of both disturbance and exchange theories and also includes the civic reform tradition and beliefs in America.[19] McFarland presents seven factors influential in the development of the public interest sector: (1) the increase in middle-class participation in American politics in the 1960s and 1970s; (2) the corresponding increase in the politics of issues and systems of beliefs as opposed to the politics of party identification, personality, or patronage; (3) the growth of "civic skepticism"—the disbelief in the utility of existing politics and public administrative practices in solving important social problems; (4) skillful leadership of public interest groups; (5) technical advances in communications; (6) economic prosperity; and (7) initial success bringing more success.[20]

McFarland's theory of representation as the balancing of civic values (even if limited to the middle class) against special interest (economically driven) values that dominate the "scope of conflict" is an accurate casting of the roles that public interest organizations have sought to play.[21] The author traces the civic balance theme to "the reform tradition of the Progressives and . . . local government reform groups" and argues that Common Cause, the focus of his later research, is simply "a recent manifestation of the mobilization of elements of the American middle class who attempt to enhance democracy and effective government through the adoption of procedural reforms supported for reasons other than clear-cut economic gain."[22] More recently, McFarland has expanded upon his argument by incorporating "political time" as a crucial element in the cyclical nature of public interest involvement.[23]

Given the civic reform tradition as well as the unique spirit of voluntarism in American society, the presence of public interest organizations as possible balancing forces is not surprising, economic research on group formation notwithstanding.[24] However, what is not as easily explained by earlier patterns of social activism is the dramatic transformation in the organization and maintenance of public interest groups during the past twenty-five years. In order to assess the changes that have taken place over the past quarter-century in the public interest sector, four data sources will be compared. The first data set was collected by Jeffrey Berry in 1972; his findings are presented in *Lobbying for the People*. More recent data on public interest organizations were collected by the Foundation for Public Affairs in 1985, 1990, and 1995.[25] From their files of more than 2,500 organizations, the Foundation staff selected 250 organizations on the basis of the following criteria:

- the extent of the group's influence on national policy,
- the number of requests received for information on the group,
- the range and quantity of news coverage generated by the group, and
- the representative nature of the group in its field of interest.[26]

Public Interest Organizations: Then and Now

One of the most obvious changes in the public interest sector over the past twenty-five years is the exponential increase in the number of groups, particularly during the period from the late 1960s to the early 1980s. Of the more than 3,000 public interest organizations in existence today less than one-third were organized three decades ago. The years of origin of the organizations studied by Berry are compared with those of the organizations presented in the Foundation for Public Affairs studies in table 1.1; these studies shall henceforth be referred to as the Berry sample and the FPA samples (1985), (1990), and (1995), respectively.

TABLE 1.1 Year of Origin: % (N)

	BERRY (1972)	FPA (1985)	FPA (1990)	FPA (1995)
1991–present	—	—	—	3.7 (9)
1986–1990	—	—	10.7 (21)	9.0 (21)
1983–1985	—	6.8 (15)	7.6 (15)	9.0 (21)
1978–1982	—	16.7 (37)	18.8 (37)	17.2 (40)
1973–1977	—	22.6 (50)	19.3 (38)	16.7 (39)
1968–1972	49.4 (39)	22.2 (49)	16.8 (33)	15.0 (35)
1960–1967	16.5 (13)	9.0 (20)	5.6 (11)	6.5 (15)
1940–1959	17.7 (14)	10.4 (23)	8.1 (16)	9.4 (22)
1920–1939	10.1 (8)	6.3 (14)	5.6 (11)	5.1 (12)
pre–1920	6.3 (5)	5.9 (13)	7.6 (15)	8.2 (19)
	(N = 79)	(N = 221)	(N = 197)	(N = 233)

While half of the organizations in the Berry sample were founded between 1968 and 1972, fully two-thirds of the FPA groups were organized after 1968. Moreover, almost half of the 1985 FPA groups and more than half
of the 1990 and 1995 FPA groups did not exist when Berry conducted his
analysis in 1972. While the organizational growth during the past three
decades is significant, a clear pattern of dramatic growth from 1968 to 1982
emerges in each of the three FPA samples. Today, more than fifteen years
after this growth spurt, approximately half of the public interest organizations operating at the national level were formed during this fertile period.
As important as the growth in the public interest sector is the ability of
many national organizations to sustain themselves for decades, despite
changing policy agendas in Washington and economic difficulties.

The Berry data offers evidence of the ability of organizations to adapt to
changes in the public interest agenda as more than 80 percent of the groups
he studied remain in operation in the 1990s. Many of the organizations no
longer in existence were directed at efforts to end the war in Southeast
Asia, clearly an issue of major import at the time of Berry's study, yet one
that no longer garners support among the citizenry in the same manner as
it did twenty-five years ago.[27] If the antiwar groups are taken out of the
sample, more than 90 percent of Berry's groups are still functioning.

With the increase in the number of public interest organizations active
today the budgets and memberships of many of these groups have also
grown. The Berry data on organizational budgets have been converted to
1990 constant dollars for an an accurate comparison of organization bud-

gets, which is presented in table 1.2.[28] By 1985 there was evident budgetary growth in public interest organizations. Almost half of the 1985 FPA organizations had budgets of $1.5 million or more, whereas only one-quarter of the Berry groups had similarly large budgets. By 1995 the growth in organizational budgets is even more pronounced as two-thirds of the FPA organizations report budgets of $1.5 million or more. In fact, 80 percent of the 1995 FPA groups have budgets of over $1 million. Even more significant, over one-quarter of the 1995 FPA organizations report budgets of more than $10 million, with four organizations over the $100 million mark. Further, it is not the case that organizations in the FPA samples are simply receiving greater support from private foundations. In fact, the general level of foundation support for public interest organizations, as reported by the major donor foundations, has declined during recent years. The findings presented in table 1.3 indicate that fewer of the FPA organizations receive almost total support from foundations, although more of them receive some level of foundation support. Overall, the level of foundation support provided to FPA organizations is approximately 20 percent of each

TABLE 1.2 Organization Budgets: % (N)

	BERRY (1972)	FPA (1985)	FPA (1990)	FPA (1995)
Less than $150,000	14.9 (11)	6.6 (13)	2.2 (4)	1.6 (3)
150,000–299,999	13.5 (10)	7.1 (14)	5.0 (9)	2.2 (4)
300,000–599,999	16.2 (12)	11.2 (22)	13.3 (24)	6.5 (12)
600,000–1,499,999	27.0 (20)	27.0 (53)	26.1 (47)	22.3 (41)
1,500,000–2,999,999	6.8 (5)	16.3 (32)	13.9 (25)	20.7 (38)
3,000,000 and above	21.6 (16)	31.6 (62)	39.4 (71)	46.7 (86)
	(N = 74)	(N = 196)	(N = 180)	(N = 184)

TABLE 1.3 Foundation Support

	BERRY (1972)	FPA (1985)	FPA (1990)
0%	55.7 (44)	42.6 (69)	36.1 (44)
less than 10%	10.1 (8)	8.0 (13)	9.0 (11)
10–49%	13.9 (11)	30.9 (50)	36.1 (44)
50–89%	8.9 (7)	16.7 (27)	17.2 (21)
90% and above	11.4 (9)	1.9 (3)	1.6 (2)
	(N = 79)	(N = 162)	(N = 122)

group's budget. Conversely, almost half of the budgets of the FPA organizations are derived from direct contributions from members or member organizations. Corporate donations, income from publications, seminars, and investments account for much of the remaining support.[29]

Related to the growth in the budgets of public interest organizations in recent years is the growth in membership. Clearly, there are more organizations with large memberships than there were two decades ago. Of those groups with memberships, almost half of the organizations in the three FPA samples have 100,000 or more members, whereas just over 20 percent of the Berry groups have similarly large memberships (see table 1.4). Since 1972 numerous organizations with widely divergent ideological perspectives have come into being, several of which have experienced dramatic growth in recent years. Considering that just two decades ago, McFarland wrote that "no conservatively oriented public interest group is now influential at the national level," the growth of such organizations is significant.[30] For example, the National Right to Life Committee, not yet organized at the national level in 1972, now claims a membership of well over one million with fifty state chapters and more than 3,000 local organizations. Founded in 1989, the Christian Coalition has grown in less than a decade to become a significant political force in American politics, with an annual budget of $20 million in 1998 and more than 1.9 million members in 1,700 chapters nationwide.

Organizations founded prior to the 1960s have also experienced dramatic increases in membership since Berry's analysis. The National Wildlife Federation had almost doubled its associate membership since 1972 to approximately one million by 1990, although its membership has dropped since that time. The Sierra Club has more than quadrupled its membership

TABLE 1.4 **Membership: % (N)**

	BERRY (1972)	FPA (1985)	FPA (1990)	FPA (1995)
no members/members =				
organizations	40.5 (32)	38.6 (59)	15.2 (19)	27.2 (41)
less than 1,000	10.1 (8)	0.7 (1)	8.0 (10)	3.3 (5)
1,000–24,999	17.7 (14)	16.3 (25)	24.0 (30)	21.2 (32)
25,000–99,999	18.9 (15)	15.0 (22)	15.2 (19)	15.2 (23)
100,000–199,999	5.1 (4)	9.2 (14)	7.2 (9)	6.6 (10)
200,000 and above	7.6 (6)	20.9 (32)	30.4 (38)	26.5 (40)
	(N = 79)	(N = 153)	(N = 125)	(N = 151)

during this time with a current membership of more than 500,000. But perhaps one of the most effective mobilizations in the public interest sector has been achieved by senior citizens groups. The American Association of Retired Persons, for example, reported just over one million members in 1968; in 1998 the AARP has more than 30 million members. More than 12 percent of the 1985 FPA groups, 14 percent of the 1990 FPA groups, and 15 percent of the 1995 FPA groups report memberships in excess of 500,000 members. Excluding the national church organizations in the Berry study (e.g., United Methodist Church, United Church of Christ, United Presbyterian Church), only one organization, the National Wildlife Foundation, had more than 500,000 members. In contrast, the forty 1995 FPA organizations with memberships greater than 200,000 have combined memberships of more than 75 *million* citizens. Controlling for organizations with no members or organizational memberships, the percentage of organizations with memberships greater than 200,000 has nearly tripled from the Berry 1972 sample to the FPA 1995 sample.[31]

As memberships have grown through the 1980s and into the early 1990s, organizations have been more likely to expand their operations by establishing chapters or branch offices. Berry reports that more than two-thirds of the organizations in his study had no chapters or branch offices. Conversely, almost half of the FPA organizations have offices or chapters beyond their national headquarters. While the bias toward having the national headquarters in the capital area remains high (well over half of the organizations in the three FPA samples are headquartered in Washington), there are several major public interest organizations that avoid moving their operations to the capital (e.g., Sierra Club, San Francisco; National Audubon Society, New York; National Association for the Advancement of Colored People, Baltimore; and Union of Concerned Scientists, Cambridge, Massachusetts). Yet, as the previous examples illustrate, most organizations not headquartered in Washington remain a short commuter flight or Amtrak ride away. More than 75 percent of the 1995 FPA groups are headquartered in the Northeast corridor from Boston to Washington.

The general pattern of expansion of operation through chapters and branch offices is even more pronounced in the growth in staffs of public interest organizations, as indicated in table 1.5. Clearly the operation of public interest organizations has become much more labor intensive over the last twenty-five years. The percentage of organizations with staffs of more than ten employees has more than tripled since 1972, (from 23 percent in the Berry sample to 75 percent in the 1995 FPA sample). And when only full-time professional staff members are considered, there has been a sev-

TABLE I.5 Organization Staffing

	OVERALL STAFF (PROFESSIONAL AND SUPPORT)		PROFESSIONAL STAFF	
	Berry (1972)	FPA (1995)	Berry (1972)	FPA (1995)
0	—	—	4.9 (4)	0.0 (0)
1	6.1 (5)	1.0 (2)	21.0 (17)	0.7 (1)
2–3	28.0 (23)	3.4 (7)	29.6 (24)	7.4 (11)
4–6	23.2 (19)	7.8 (16)	19.8 (16)	10.8 (16)
7–10	19.5 (16)	12.7 (26)	16.0 (13)	21.0 (31)
11–20	12.2 (10)	18.6 (38)	4.9 (4)	22.3 (33)
21–40	6.1 (5)	18.6 (38)	1.2 (1)	15.5 (23)
40+	4.9 (4)	37.9 (78)	2.5 (2)	22.3 (33)
	(N = 82)	(N = 205)	(N = 81)	(N = 148)

en-fold increase in the percentage of groups with more than ten professionals (from under 9 percent in the Berry sample to more than 60 percent
in the 1995 FPA sample).

While there are indeed more public interest organizations active in the
political arena today than at any time in our nation's history, contemporary
groups have quickly adapted to the increasingly complex political environment. With significant growth in memberships, budgets, and staffs over
the last two decades, today's public interest organizations are quite distinct
from their predecessors. The organizational transformations within the
public interest sector have resulted in a more professionalized and increasingly competitive political marketplace. In the last few years this competition for resources has forced many public interest organizations to begin to
downsize and has forced mergers of similar organizations, not unlike the
developments in corporate America.

A significant factor in competing for members is the tax status of public
interest organizations. Tax exemption is widely viewed as a valuable asset to
organizations as it provides exemption from business taxation and provides
significantly reduced mailing rates, crucial to direct mail recruitment efforts. The public interest sector today contains three basic types of organizations defined by the Internal Revenue Service: (1) 501(c)(3) organizations,
(2) 501(c)(4) organizations, and (3) 501(c)(4) organizations with affiliated
501(c)(3) foundations.[32] The first type, 501(c)(3) groups, allows supporters
(members) to deduct their contributions on their income tax returns. But in
order for an organization to receive tax deductibility for members, it is pro-

hibited from devoting "a substantial part of its activities" to "carrying on propaganda or otherwise attempting to influence legislation."

In its original form, as a part of the Revenue Act of 1934 (Code revised in 1954), the statute made no effort to define "propaganda," "legislation," or "substantial part." Since its inception, Congress and the courts have managed to provide some concrete parameters to this regulation. Congress, in the Tax Reform Act of 1976, provided alternative tests for public charities in order to bypass the original "substantial part" guidelines and denied absolutely the right of organizations to "participate in, or intervene in, any political campaign on behalf of any candidate for public office." Federal courts in *Haswell v. U.S.* 500 F2d 1133 (1974) have clarified the issue by defining which activities did not constitute lobbying activities. The court ruled that "technical advice or assistance" and "nonpartisan analysis, study, or research" does not constitute lobbying only when the organization receives a "specific written request" for such information. The final rule-making process relating to the 1976 act in the IRS concluded with a set of regulations issued fourteen years later in August of 1990; these regulations listed additional legitimate political activities, such as the presentation of information to Congress (without request) when the information is also available to the general public and nonpartisan voter registration activities (carried on in five or more states).[33]

However, 501(c)(4) organizations need not meet these requirements, i.e., they are able to lobby Congress and the executive branch on a full-time basis. They do sacrifice the ability of their members to deduct their contributions as a result. As Berry notes, the "most advantageous alternative" is what has more recently been called "piggybacking," the structuring of the organization in such a way that there are two distinct entities: a lobbying and influence enterprise (501[c][4]) and an education-research wing (501[c][3]). As one exempt organizations specialist explains, "The (c)(3) is the idea chamber, and the (c)(4) is the money chamber that puts the idea on the street."[34] Thus, 501(c)(3) foundations affiliated with 501(c)(4) groups may also become depositories for organizational costs. Overhead expenses, salaries, rent payments, etc. are easily shuffled from one organization to the other. In fact, there are instances where direct transfers of 501(c)(3) funds are made to the 501(c)(4) parent organization. For example, according to the 1985 financial reports of Environmental Action, Inc. (501[c][4]), and Environmental Action Foundation (501[c][3]), 15 percent of EAF's total expenses ($52,300 of $333,541) appear in the form of a grant to EA, Inc.—tax-deductible funds transferred to an organization that is not tax-deductible.[35]

Financial information on all public interest organizations is, at least theoretically, available through the Internal Revenue Service. Nonprofit organizations are required by law to file annual reports with the IRS, providing full financial disclosure of organizational activities (known as 990 forms). However, a 1983 General Accounting Office study of approximately 11,000 501(c)(3) organizations revealed that 94 percent "did not completely respond to certain public information reporting requirements."[36] Unfortunately the IRS does not have the capacity to address these reporting discrepancies. In 1996 there were approximately 1.1 million tax-exempt organizations reporting organizational revenues of more than *$1 trillion* in total, not including revenues generated by churches across America. Today, the IRS division responsible for regulating these nonprofit organizations employs less than 1,000 workers.[37] While there are instances in which tax-exempt organizations have relinquished their 501(c)(3) status due to excessive political activity, (e.g., Sierra Club in 1966), many organizations, up until quite recently, have managed to escape the wrath of the IRS while lobbying Congress beyond allowable limits. For the past several years, however, attention has been focused on the nonprofit sector and particularly on public interest organizations. The shift in philosophical orientation in Congress, due to the Republican successes in the 1994 congressional elections and the retention of the Republican majority in Congress since that time, has resulted in the close scrutiny of this government-subsidized nonprofit sector of society.

With the passage of the Lobbying Disclosure Act of 1995, the act of lobbying has been expanded to include not only direct contacts with members of Congress, but also contacts with congressional staff as well as with political appointees of the executive branch (e.g., cabinet secretaries, agency administrators) and with senior members of the executive bureaucracy.[38] To appease the nonprofit sector, the law gives nonprofits the option of following existing IRS guidelines regarding the definition of lobbying, which does not include contacts with staff or the executive branch, but does include grassroots efforts, or of abiding by the new definition of lobbying in the 1995 act. The nonprofit sector, led by the Independent Sector and the Alliance for Justice, was instrumental in killing the lobby reform effort in the 103rd Congress because the language in that legislation would not have given nonprofits the option of using the IRS guidelines.[39]

There is one provision, however, in the Lobbying Disclosure Act of 1995 that does affect certain 501(c)(4) organizations. Known as the Simpson Amendment prior to final passage, section 18 of the new act prohibits from lobbying any 501(c)(4) organization that receives federal funds in the form of grants, contracts, or loans. Senator Alan Simpson (R-WY) had the American Association of Retired Persons (AARP) in mind when he drafted his amendment. In 1995 Simpson conducted a series of hearings that focused on the internal operations of the AARP. The senator was concerned with the intermixing of for-profit and nonprofit activities in the organization. While the hearings were inconclusive, Simpson went forward with his amendment; it was subsequently accepted by both houses and included in the final act.[40]

Similar efforts to restrain the activities of public interest organizations are occurring in the House of Representatives. Representatives Ernest Jim Istook (R-Okla) and David McIntosh (R-Ind) seek to ban all nonprofits that receive government funds from lobbying. Their measure was proposed as an amendment to the Lobbying Disclosure Act as well as to a host of unrelated bills, but it was unsuccessful in the 104th and 105th Congresses.[41] The chief target of Istook and McIntosh is the National Council of Senior Citizens, a liberal senior citizens organization aligned with the labor movement that receives more than 80 percent of its revenues from the U.S. Department of Labor. Nevertheless, all public interest organizations are concerned about the Istook-McIntosh proposal.

While the sheer magnitude of change in the organization of public interest enterprises during the past twenty-five years is quite evident, there remains one organizational aspect that has not undergone such dramatic change—the composition of organizational leadership. Executive control over public interest organizations through executive directorships and governing boards remains overwhelmingly in the hands of white males. Men occupy 75 to 80 percent of public interest executive directorships, according to the 1985 and 1990 FPA samples (see table 1.6). The organizations led by women tend to have smaller organizational budgets and also tend to be more recently organized rather than older, established organizations.

TABLE 1.6 Executive Directorships by Organizational Budgets

BUDGET SIZE	1985 FPA	1990 FPA
	*Female % (N)**	*Female % (N)*
under $1,500,000	31.9 (94)	35.4 (82)
1,500,000–$3,000,000	14.6 (41)	28.0 (25)
over $3,000,000	7.0 (86)	18.3 (71)

*N = Total number of organizations in each budgetary category.

Similar patterns are found when analyzing the composition of the governing boards of public interest organizations. Indeed, organizations with female executive directors tend to have significantly greater representation of women on their governing boards (see table 1.7). Overall less than one-quarter of public interest board positions are occupied by women. Service on public interest boards is limited by virtue of the multiple qualifications necessary: managerial skills, policy expertise, financial input, political connections, and celebrity status. Individuals possessing some useful combination of these attributes make strong candidates for board service. In addition to qualifications, there is a significant amount of insularity within and among these boards.

Political credentials are quite useful for entrance into the Washington public interest community, which in turn creates access to the institutions of governance. An analysis of the top twenty-five officials for the first term of the Clinton administration shows that only Education Secretary Richard Riley, former governor of South Carolina, former Transportation Secretary and Secretary of Energy Federico Pena, Attorney General Janet Reno, and EPA Administrator Carol Browner are not represented, directly or indirectly, in the public interest leadership community.[42] For example, the first lady, Hillary Clinton, has served as chair of the board of the Children's Defense Fund, while Health and Human Services Secretary Donna

TABLE 1.7 **Board Representation in Public Interest Organizations**
      (1985 FPA Sample)

% FEMALE BOARD MEMBERS	% OF ORGANIZATIONS	% FEMALE EXECUTIVE DIRECTORS
0%	21.1% (N = 38)	0.0%
1–20%	32.2% (N = 58)	15.8%
21–40%	30.6% (N = 55)	21.8%
41–60%	9.4% (N = 17)	37.5%
61–80%	2.8% (N = 5)	60.0%
81–100%	3.9% (N = 7)[a]	85.7%
	Sample Mean 22.7%	
	(N = 180)	

[a]Organizations with more than 80 percent female representation on boards include the Gray Panthers, National Abortion Rights Action League, Women's Equity Action League, Women's Legal Defense Fund, Eagle Forum, League of Women's Voters, and the National Women's Political Caucus.

Shalala served as its vice-chair. Tipper Gore, wife of Vice President Al Gore, founded the Parent's Music Resource Center; former Treasury Secretary Lloyd Bentsen's wife, B.A., serves on this board as well. Interior Secretary Bruce Babbitt was formerly president of the League of Conservation Voters. Robert Reich, labor secretary during Clinton's first term, served on the boards of the Economic Policy Institute and the Business Enterprise Trust. Hazel O'Leary, energy secretary during the first term, served as vice-chair of the board of the Keystone Center. Secretary of Housing and Urban Development Henry Cisneros served on the board of the National Council on the Aging. Jesse Brown, veterans affairs secretary, was formerly the director of the Disabled American Veterans organization. U.S. Representative to the United Nations and current Secretary of State Madeleine Albright presided over the Center for National Policy, while Treasury Secretary Robert Rubin served on its board. Former Director of the Office of Management and Budget and current vice-chair of the Federal Reserve Board Alice Rivlin was chairperson of the board of The Wilderness Society, and Warren Christopher, secretary of state during Clinton's first term was vice-chair of the Council on Foreign Relations.[43] Mickey Kantor, former U.S. trade representative and Clinton's second commerce secretary, served with Hillary Clinton on the board of the Legal Services Corporation. When former members of Congress who have served in the administration are included, as well as the late Commerce Secretary Ron Brown, who formerly presided over the Urban League and the Democratic National Committee, it is evident that the Clinton White House has demonstrable ties to the Washington public interest political community. Interestingly, approximately one-quarter of the leadership corps assembled by the president is composed of women, mirroring the composition of the public interest leadership community.[44]

Theoretical Linkages

The preceding substantive overview outlines the scope of the significant organizational change that has occurred over the last twenty-five years in the public interest sector. In similar fashion, the scholarly study of interest group activism in general, and the role of public interest organizations in particular, has produced a wide variety of theoretical and analytical perspectives that incorporate the organizational dynamics of the past quarter-century. In addition to the works of Truman, Salisbury, Berry, Schlozman and Tierney, and McFarland presented earlier, numerous scholars have shaped the research that will follow in subsequent chapters. I will summarize briefly the major scholarly antecedents in the hope that the interested reader will study these works as well.

One cannot discuss the organizational aspects of political organizations without acknowledging the research of the late economist Mancur Olson. The economic model of individual behavior posited by Olson assumes that citizens behave in an economically rational manner; such an assumption leaves little room for the existence of public interest organizations. Olson argued that it is not economically rational to join an organization that provides collective benefits, i.e., consumer protection, clean air, racial justice, or public safety. Economically rational citizens would elect to get a "free ride" or receive the benefits of the efforts of public interest organizations without paying the organizational costs of membership dues necessary to join such an organization.[45]

Political scientist Terry Moe and sociologist David Knoke, while acknowledging the validity of the claims made by Olson, sought to expand his argument beyond the scope of the narrowly defined economic rational actor model. Each scholar has incorporated additional incentives that may cause citizens to overcome the free-rider predisposition in order to join public interest organizations. The analysis presented in chapter 5 draws heavily upon the research of Moe and Knoke.[46] In addition to these incentive theorists, I have incorporated key components of the resource mobilization perspective, first articulated by John McCarthy and Mayer Zald and employed widely in sociological research, and have linked this perspective to the early work of Theodore Lowi, built upon by McFarland, that offers a developmental rationale explaining the transformation of the social movements of the 1960s into the professionalized public interest organizations of the 1990s.[47] McCarthy and Zald summarize the resource mobilization perspective as follows:

> The resource mobilization approach emphasizes both societal support and constraint of social movement phenomena. It examines the variety of resources that must be mobilized, the linkages of social movements to other groups, the dependence of movements upon external support for success, and the tactics used by authorities to control or incorporate movements. . . .
>
> The new approach depends more upon political, sociological, and economic theories than upon the social psychology of collective behavior.[48]

This interdisciplinary mix of scholarly research allows for a more inclusive analysis of organizational attributes, the relationships between organizational leaders and their members, and provides an operational definition of public interest organizations to be used in the chapters that follow. Incorporating the various perspectives presented above as well as earlier in the chapter, *a public interest organization is defined as a complex, institutionalized, professionalized 501(c)(3) or 501(c)(4) nonprofit organization that seeks to represent collective interests in American society to federal policy makers in a direct or indirect manner, the representation of which will not selectively benefit the leadership or membership of the organization.*

Organizational Focus

The definition of public interest organization presented above allows for a broad range of political organizations to be included, traditional public interest organizations, such as Common Cause, Public Citizen, and the American Civil Liberties Union, as well as environmental groups, consumer groups, civil rights groups, peace groups, and taxpayers organizations. In addition, single-issue groups meet the definitional requirements. Pro-life and pro-choice groups organized at the national level are included; likewise, the National Rifle Association is a public interest organization. At the same time, however, there are restrictions implied in the definition. Public interest organizations, as defined in this study, are organized at the national level to represent broad societal interests. While not precluding the organizations from attempting to influence the policy process at the state, local, or international levels, organizations presented in this study must be organized with a national focus and include broad membership representation. Therefore, the analysis that follows explicitly does not apply to local grassroots organizations or to international nongovernmental organizations.

There are several reasons for such an organizational limitation. First, the organizational structures of local groups are rarely as bureaucratic or elaborate as those found at the national level. Second, staffing is largely voluntary in grassroots groups, whereas the data presented earlier in the chapter demonstrate the level of professional staffing in national organizations. Third, the relationships between leaders and members, including initial recruitment and retention, are quite different at the local and national levels. Fourth, internal relationships between staffs and governing boards are not at all similar at these two levels. Support for these distinctions is found in the excellent research published by Laura Woliver and by Jeffrey Berry, Kent Portney, and Ken Thomson.[49] Interested readers will find stark contrasts between the analysis presented in this study and the findings of these scholars who focused on organizational activism at the local grassroots level.

In terms of substance, the organizational transformations that have occurred in the environmental movement at the national level during the past three decades are so dramatic and diverse that focusing on this sector of the public interest movement will yield lessons that are applicable not only to national environmental organizations planning for the next century but also to the broader set of public interest organizations operating at the national level.

As a result, in the chapters that follow, only national environmental organizations will be analyzed. While local NIMBY (Not In My Back Yard) groups, grassroots movements, and loosely organized direct action "ecotage" groups that operate outside of the legal-political system will be discussed in the context of environmental activism in the 1990s, the findings presented in subsequent chapters are not applicable to such local grassroots groups. As the contemporary public interest sector, and particularly the national environmental component, continues to adapt to the changing political and economic climate, there is one key assumption upon which this research is based, namely, that nationally organized public interest organizations will endure well into the next century. While their structures may change and their institutional focus may shift, due to competing forces of localization (through state, grassroots, and NIMBY groups)[50] and globalization (through international nongovernmental organizations),[51] such organizations will remain significant actors in the national political arena.

But before assessing the organizational state of affairs of national environmental organizations, it is necessary to underscore the tensions that exist between maintaining the daily operations of these national organizations and representing the political interests of the organizations in the policy-making process. As a result of the rapid growth and subsequent organizational transformation in the public interest sector in the 1980s and 1990s, public interest leaders as well as their citizen supporters are confronted with a new set of concerns. The issues currently facing public interest organizations are somewhat different from what Jo Freeman labeled two decades ago the "classic dilemma of social movement organizations: the fact that the tightly organized, hierarchical structure necessary to change social institutions conflicts directly with the participatory style necessary to maintain membership support and the democratic nature of the movement's goals."[52]

While the current relationship between organizational leaders and their members focuses on the issue of organizational maintenance and political representation, broadly defined, Freeman's dilemma is no longer an issue in the public interest sector. Even in the most structurally representative organizations, bureaucratization of the leadership infrastructure is an organizational fact of life. The contemporary relationship between organizational maintenance and political representation in the public interest sector may be easily misinterpreted if organizational institutionalization is the singular focus. Terry Moe notes that "the degree of internal democracy can be treated, from a pluralist standpoint, as something that does not really affect the more fundamental congruence between group goals and member goals—since congruence is guaranteed by the pluralist logic of membership. If individuals join (quit) interest groups because they agree (disagree) with their policies, then the question of oligarchy is in some sense beside the point."[53]

While the bureaucratic structuring of public interest organizations has a direct impact on the quality of political representation, Freeman's dilemma includes the implicit assumption that public interest representation is always the primary product of public interest organizations. Today, such an assumption should not be made. Public interest leaders are now faced with the primary task of keeping their organizations in operation and the secondary task of being effective public interest advocates. As these organizations have developed, the costs of maintaining their day-to-day operations and, more important, of maintaining membership bases in the increasingly competitive public interest marketplace have markedly shifted organizational resources toward maintenance of the organization and, as a consequence, away from public interest representation.

Membership retention in the public interest sector in the 1990s, according to David Cohen, often amounts to a quasi-economic transaction: "People do vote with their checkbooks. That is one form of participation. But just as organizations cannot thrive solely through checkbook participation, neither can they survive the absence of dues payers and voluntary contributors. Otherwise there is no organizational accountability for people to have a voice or have the chance of building loyalty."[54] As a result, the relationship between organizational maintenance and political representation in the public interest sector at the end of the 1990s is increasingly complex. Public interest leaders are faced with difficult choices regarding the allocation of resources supplied, in large part, by members committed to public interest goals rather than to the maintenance of public interest organizations.

Finally, it is necessary to give a clearer meaning to the concept of political representation as it applies to the public interest sector. Political representation, at least in democratic societies, assumes that a significant linkage exists between leaders and constituents. The republican form of government practiced in the United States is predicated on the notion that governing institutions and policy will reflect the wishes of the governed populace. Political theorist Hannah Pitkin offers an excellent working definition of political representation, which is adaptable to the special conditions of public interest representation.

> Political representation is primarily a public, institutionalized arrangement involving many people and groups, and operating in the complex ways of large-scale social arrangements. What makes it representation is not any specific action by any one participant, but the overall structure and functioning of the system, the patterns emerging from the multiple activities of many people. It is representation if the people (or a constituency) are present in governmental action, even though they do not literally act for themselves.[55]

Public interest representation differs only in the degree to which "the people (or a constituency)" actually participate in policy influence. The representatives in this context are not governmental policy makers; rather they are individual constituents and organization leaders acting to influence the policy makers. In essence, public interest representation is a phenomenon on the subsystem level if one considers the representative relationship between those who govern and the governed as the political system within which public interest organizations and their adherents operate. Throughout this study, political representation in the public interest sector will be analyzed in the context of this overarching political environment.

Political representation may be evaluated based on several criteria. In the chapters that follow, this concept is analyzed from several perspectives in order to identify its various manifestations. Organizational structures are analyzed to determine the representational consequences of distinct organizational forms. The motivational component of public interest representation, including the theoretical assumptions surrounding the motivations of individuals for joining and remaining with public interest organizations, is explored as well. In addition, the activities of individuals, as active citizens and organization advocates, highlight the special relationships between organizational leaders and members and the policy makers in Washington. Finally, representation may be measured as political influence—organizational effectiveness in the governmental policy-making process.

Together, these three dimensions of public interest representation constitute a new type of political representation—representation as "acting with." Public interest representation is a mixture of "actual" and "virtual" representation at the subsystem level. It is actual representation in the sense that active citizens, operating in the noncoercive context of public interest organizations, are free to join as well as exit the organizations and are also free to act as individual influence agents through direct contacts with policy makers.[56] Public interest representation is virtual to the extent that organization leaders are "acting for" their members as well as for the larger societal interests.[57] Both individuals (members) and organization leaders are free to act on their own initiative. Nonetheless, effective political representation in the public interest sector is more easily achieved when these two components are acting in concert, when the organizational *voices* are boldly *echoed* by their members. It is within this political context that the efforts at organizational maintenance and political representation within and among national environmental organizations will be explored.

More than any other movement in the public interest sector during the past three decades, the environmental movement has undergone dramatic changes, not only in its organizational composition, but also in the means of articulating environmental interests in the policy-making process. Compared with the organizational and representational methods of the environmental movement of the late 1960s, the concepts of organizational maintenance and political representation in the late 1990s have very different meanings for the vast majority of national environmental organizations.

In the next chapter, the transformation in the organization of American environmentalism from a broad, loosely defined social movement to a collection of professionalized public interest organizations is explored. Organizational data from a broad variety of environmental groups will be presented and evaluated in the context of the changes in societal attitudes toward the environment, as measured by national public opinion surveys and by media coverage during the past twenty-five years. Following the broad overview of the transformation of the environmental movement, five specific environmental groups will be presented: Environmental Action, Environmental Defense Fund, National Wildlife Federation, Sierra Club, and The Wilderness Society. These organizations will serve as a representative cross section of the groups active in the national environmental movement during the past three decades, as each offers diverse sets of organizational, representational, and ideological perspectives and has undergone various successes and failures in organizational management and leadership, including the demise of one of the organizations. Chapters 3, 4, 5, and 6 will focus on the various organizational and representational dimensions of these five groups but will also include relevant examples from other environmental organizations.

From Social Movement to Public Interest Organizations

The Organizational Transformation of Environmentalism in the United States

Professionals keep the movement organized.
Amateurs keep it honest.   —STEPHEN FOX, *John Muir and His Legacy*

In no other arena of the public interest nonprofit sector has the transformation from social movement to contemporary professionalized national organizations been as pronounced and significant as in the environmental movement.[1] Forty years ago, the environmental movement in the United States included about 150,000 citizens who belonged to environmental groups. The collective financial resources of existing groups amounted to less than $20 million annually. Today, more than eight million citizens are affiliated with national environmental organizations, contributing more than $750 million a year to environmental causes. By the end of the century, the collective wealth of the national environmental movement will likely reach the $1 billion mark.

Prior to the 1960s, the emergence and organizational development of the American conservation movement in the United States was an important antecedent to the more recent environmental expansion. In many historical accounts, special attention is given to the emergence of organized constituencies focused on conservation of the nation's natural resources.[2] The earliest citizen mobilizations for the environment appeared as conservation groups in the late nineteenth century; however, adherents to the cause of conservation were sufficiently diverse, from the progressive "wise use" followers at the turn of the century to wilderness preservationists, that the movement, such as it was, remained unorganized, "small, divided, and frequently uncertain."[3] The Sierra Club, founded in 1892, the National Audubon Society, founded in 1905, and the National Parks and Conservation Association, founded in 1919, are three of the oldest organizations in the contemporary environmental movement. During the first half-century of the conservation movement, none of these organizations was developed to the extent each is today. While both the Sierra Club and the National Audubon Society have memberships of more than half a million people, the bulk of this growth has occurred after the substantive shift in the movement from conservationism to environmentalism, that is, since the mid-1960s. For example, the Sierra Club membership rolls did not reach 1,000 individuals until 1908; by 1930 its membership finally exceeded 2,500. Even in 1960, only 16,000 citizens claimed affiliation with the Sierra Club, and most of them resided in California.[4]

Despite the slow growth in older conservation organizations during the first half of the twentieth century, between the two world wars, several more national organizations were established in pursuit of broadly defined conservation goals. Two organizations, the Izaak Walton League and the National Wildlife Federation, formed in 1922 and 1936, respectively, were founded by sportsmen and supported conservation efforts in order to preserve natural areas for sporting purposes. Today, the sportsmen component of these organizations still exists; however, both organizations have broadened their agendas considerably. A third interwar organization, The Wilderness Society, was formed in 1935 in order to draw attention to the plight of wilderness areas not incorporated into the National Park system in the United States.[5]

With the exception of the creation of Defenders of Wildlife in 1959, the conservation movement remained a small band of loosely organized individuals. At most the conservation movement, as it existed in 1960, included only about 150,000 active members across the country. Since that time a dramatic transformation in the environmental and conservation arena has taken place. Particularly important in this transformation "was the shift from the complacency of the 1950s to the rising social consciousness of the 1960s with its willingness to acknowledge, and to challenge, the darker side of the American dream."[6] Changes in society itself placed conservationists in a position of becoming marginalized by a growing, qualitatively different set of "environmental" issues. The effects of massive postwar development on the nation's natural resources began to be felt. New technologies, readily accepted decades earlier, were now being closely scrutinized for their potential negative side effects, both on human beings and on nature. By the mid-1960s a wide variety of scholars, authors, and activists, including Rachel Carson, Barry Commoner, Paul Ehrlich, Garrett Hardin, and Dirck Van Sickle, had published treatises on the harmful effects of environmental pollution, pesticides, overpopulation, and nuclear power.[7]

Together these new issues, concerns, and attitudes provided the necessary preconditions for the emergence of the new environmental era.[8] The transformation from conservation movement to environmental movement occurred in the late 1960s, with the creation of several new environmental organizations. Activated by the publication of *Silent Spring* by Rachel Carson, the first of the new breed of environmental organizations, the Environmental Defense Fund, was created in 1967 by a group of concerned scientists in an effort to battle the use of DDT on Long Island, New York. Through the assistance of grants from major foundations as well as through contributions from other major donors in the private sector, organizations such as the Environmental Defense Fund, the Natural Resources Defense Council, Environmental Action, and Friends of the Earth were established, several in conjunction with the celebration of the first Earth Day in April 1970.[9] With new environmental agendas, new resources, and new strategies of influence (e.g., litigation), these organizations sought to redefine the concept of environmentalism for the American public.

Older, more established conservation organizations took note of the new brand of environmental advocacy being practiced. While the issue agendas of these organizations still include the traditional wilderness preservation and conservation emphasis, even they have adopted many of the "environmental" concerns identified by newer organizations. Furthermore, they have adopted the more professionalized approach to organizational management and environmental advocacy demonstrated by their new counterparts in the movement.

By the mid-1980s, both old-line conservation groups and the newer environmental organizations had professionals serving in virtually all organizational capacities, not only in direct policy advocacy. Development specialists organized high-technology direct-mail and telemarketing efforts to recruit and maintain increasingly volatile memberships. Personnel managers conducted the operations of these organizations in a businesslike manner. From their executive directors to low level staffers, environmental organizations shed their amateur structure and image for a more professional look. Environmental leaders developed a bottom-line approach to the governance of environmental organizations. According to Grant Thompson, senior associate at the Conservation Foundation:

> The sudden embrace of business attitudes arises out of a need to marshall resources effectively. In the business world, dollars are the measure of value added and profits are the measure of effectiveness. In the nonprofit world, the measures are less clear, and this has led some to the mis-

taken notion that dollars do not matter. Yet in the nonprofit world, as surely as in the for-profit one, dollars quantify the resources that are available to the organization for carrying out its mission.[10]

Environmental Defense Fund staff lawyer David Roe also stressed the inherent importance of organizational maintenance: "Public interest work is nonprofit, inevitably and proudly; but it is hardly immune from economics. Instead of profit, what measures economic success is the support of donors."[11]

This dramatic organizational restructuring was facilitated by the significant influx of financial resources generated by a new wave of environmental mobilization in the 1980s. Two key organizational and political factors help explain the exponential growth in the environmental movement in that decade. From no more than 150,000 followers and a total budget of less than $20 million, the environmental movement had grown by the end of the 1980s to more than 100 national organizations (and more than 10,000 state and local groups) with a combined membership of more than eight million. The combined budgets of these organizations approached $500 million, representing more than a twentyfold budget increase in less than thirty years.[12]

Clearly, the key political factor was Ronald Reagan. The policies of the Reagan administration served as catalysts for dramatic growth in the organized environmental movement as well as in other organizations championing liberal causes. More important, these policies were personified not only in Ronald Reagan himself, but in the political appointees selected by the Reagan administration, particularly in the environmental policy arena. The selections of James Watt as secretary of interior and Anne Gorsuch Burford as administrator of the Environmental Protection Agency provided environmental organizations with *real* public enemies.[13] To this day, veteran staffers in the environmental community still refer to a significant cohort of members who joined their organizations in the early 1980s as "Watt babies."

Beyond an identifiable enemy, the key organizational factor was the ability of environmental organizations to market their messages to increasingly wider audiences. Certainly the shift from social network recruiting (friends recruiting friends) to direct-mail solicitation efforts generated a much larger constituency of interest for the growing number of organizations in the environmental arena. Thanks to the confluence of these two factors, the rapid expansion and resulting restructuring of the movement into a series of competing political enterprises was completed in relatively short order. From an organizational perspective, the decade of the 1980s was literally a "boom time" for most national environmental organizations. Budgets and memberships typically doubled or tripled during the decade.[14]

Unfortunately for a number of environmental groups, the organizational boom of the 1980s ended with the decade, forcing many groups to suffer through the growing pains associated with rapid budgetary expansion and organizational transformation followed by flat or declining budgets and decreasing memberships. These pains became acute in the first few years of the 1990s. For many the national environmental organizations, the era of organizational prosperity had come to an abrupt end. After years of budgetary growth in the late 1980s, with average annual increases in donations running at 20 percent for the ten largest organizations, many organizations experienced revenue shortfalls and significant declines in memberships. From 1990 to 1993 memberships in the ten largest groups declined by more than 500,000, a 6 percent aggregate decrease. Many organizational budgets were flat or actually decreased from 1990 to 1994. This financial turnaround was particularly painful for some groups, given the period of rapid organizational expansion in the mid-1980s. Staff layoffs occurred in a number of organizations as well as cutbacks in programs and office closings. Leaders were not immune to the downsizing as several long-time presiding officers of the most high-profile environmental organizations were removed amid allegations of mismanagement.

Today, most environmental organizations have changed in response to the organizational turmoil of the early 1990s, while others have managed to weather the storm relatively unscathed. In fact, a number of organizations, including the Environmental Defense Fund, the National Parks and Conservation Association, and the Nature Conservancy, experienced significant growth in the first half of the 1990s.[15] By the end of 1995, the combined budgets of national environmental organizations surpassed $700 million, a net increase of 40 percent since 1990.[16]

In the chapters that follow, the reasons for the organization and membership volatility within national environmental organizations during this period will be analyzed by examining the impact of growth in the 1980s on organizational maintenance and representation in the 1990s. But first it is necessary to establish the larger political context within which the environmental movement has developed during the past two decades.

Public Opinion, the Media, and the Environment

"Ask Americans if they favor environmental protection and the vast majority answer, 'Of course!' Poll after poll in recent years confirms this."[17] Journalist Brad Knickerbocker is correct in his general assessment of American public opinion toward the environment. Recent national polls do reflect overwhelming levels of support or sympathy for environmental concerns (see table 2.1). Consistently more than three-quarters of those surveyed in national polls either identify themselves as "environmentalists" or are sympathetic toward the cause. These findings, while widely reported in media accounts, often mask the true relationships between citizens and the environment. Everett Carll Ladd and Karlyn H. Bowman, in their recent analysis of public attitudes toward the environment over the past twenty-five years, argue that the broad patterns of support for the environment are not properly placed in the context of other important societal issues, such as economic growth and government regulation:

> As we reviewed hundreds of polling questions asked about the environment since 1970, we have been reminded of one central weakness of the huge collection of attitudinal data on the subject. . . . Americans have been asked repeatedly in a wide variety of formulations to affirm a core value, in this case the importance of the environment. Each time, not surprisingly, they responded that a clean and healthful environment was important to them. These questions tell us little about what a society with many demands on it is willing to do to advance the value, or what happens when one important value clashes with another.[18]

The decade of the 1980s was a time of relative economic prosperity. Inflation and unemployment rates were much lower than during the Carter

---

TABLE 2.1 **Views on the Environment**

Question: In recent years, interest in the environment has increased. Do you think of yourself as an active environmentalist, sympathetic toward environmental concerns but not active, or neutral, or generally unsympathetic to environmental concerns?

	1992	1993	1994	1995
Active environmentalist	29	21	23	21
Sympathetic, but not active	52	55	56	54
Neutral	15	20	16	19
Unsympathetic	2	2	2	4

Sources: Surveys by Roper Starch Worldwide, Inc. for Times-Mirror Magazines; presented in Ladd and Bowman, *Attitudes Toward the Environment,* table 4, p. 15; 1995 data direct from Roper Starch Worldwide, Inc.

presidency of the late 1970s. Citizens' concerns were not exclusively focused on the economy. As a result, the clashes of values presented by Ladd and Bowman above were not as clearly articulated. In the 1980s roughly one-third of citizens, in response to public opinion surveys, were more likely to view environmental protection as a value for which they were willing to sacrifice economic growth; about half of the American public felt that our society could experience economic growth while also providing for a quality environment. The gap between these to groups of citizens was approximately twenty percentage points.

Clearly by the 1992 presidential election, however, the state of the economy was a primary concern for the vast majority of Americans. Consequently, issues such as preserving a clean environment were given less consideration. More important, as the economy became the paramount political issue, growing numbers of citizens were willing to view environmental protection and economic growth as complementary goals (see table 2.2). By 1994 what had been a twenty-point gap between pro-growth environmentalists and anti-growth environmentalists had doubled to more than forty percentage points, with two-thirds of citizens adhering to the pro-growth environmental position and only one-quarter finding economic growth and environmental quality to be incompatible.

Similarly, throughout the 1980s, a majority of Americans believed that the federal government had too little involvement in providing for environmental protection. On average only 10 percent of the population believed that the federal government was too involved in regulating society to protect the environment (see table 2.3). The gap between those reporting too little involvement and those reporting too much involvement was, on average, 45 percentage points from 1984 to 1991. Between 1991 and 1994 the relationship changed dramatically. While the percentage of citizens believing there was too little governmental involvement in environmental protection dropped by fifteen points, the percentage of those claiming too much governmental involvement has grown by twenty points. Between 1992 and 1994 the gap between these opposing positions was cut in half, with only a nine-point difference in the 1994 sample. Similar findings were reported in a 1996 national survey conducted for the Superfund Reform Coalition.

Beyond these broad societal views captured in public opinion polls, two additional resources are available to assess the political context within which national environmental organizations operate. The first, while also consisting of polling data, provides a measure of salience of environmental concerns in the context of voting. Virtually all of the major media outlets in

TABLE 2.2 Economic Growth versus Environmental Protection

Question: Which of the two statements is closer to your opinion? There is no relationship between economic growth and the quality of the environment—indeed, we can have more and more goods and services and also a clean world. OR: We cannot have both economic growth and a high level of environmental quality; we must sacrifice one or the other.

	1980	81	82	83	84	85	86	87	88	89	90	91	92	93	94
Can have growth and clean world	50%	45	49	49	55	51	51	56	49	49	59	61	68	67	67
Cannot have both	31%	35	35	27	28	33	31	31	31	33	32	31	23	25	25

Sources: Surveys by Cambridge Reports/Research International, presented in Ladd and Bowman, *Attitudes Toward the Environment*, table 1, p. 9.

TABLE 2.3 Government Regulation and Involvement in the Environment

Question: In general, do you think there is too much, too little, or about the right amount of government regulation and involvement in the area of environmental protection?

	1984	1985	1986	1987	1988	1989	1990	1991	1992	1993	1994	1996
Too little involvement	56	54	59	49	53	58	62	60	55	49	40	36[a]
About the right amount	27	28	26	30	25	24	15	16	16	21	22	N/A
Too much involvement	8	10	7	12	12	9	16	16	22	25	31	21[a]

Sources: Surveys by Cambridge Reports/Research International, presented in Ladd and Bowman, *Attitudes Toward the Environment,* table 9, p. 23; Linda DiVall, Superfund Reform Coalition Survey, 1996.

[a]DiVall's survey reported 21 percent of respondents agreeing that "environmental laws have gone too far," while 36 percent believed that current "environmental laws have not gone far enough."

the United States conduct exit polls or pool their resources and conduct joint efforts (e.g., Voter News Service). Upon leaving the voting location, voters across the nation are asked a series of questions regarding their voting choices and the issues they regard as important in their voting decisions.

The exit polling results presented in table 2.4 provide comparisons between the most important issue chosen by voters in a variety of elections and the environment as the most important issue in those elections. Since 1980 the environment has appeared among the top four election issues only once, in 1990 after the passage of the amendments to the Clean Air Act, when approximately one-fifth of voters interviewed identified the environ-

TABLE 2.4 **The Environment as a Voting Issue**

Date	Exit Pollster	Most Important Issue (Compared with the Environment)		Environmental Voters
1982	CBS/New York Times	Unemployment	38%	
		Environment	3%	
1984	Los Angeles Times	Govt. Spending	22%	
		Environment	4%	Mondale 75% Reagan 25%
1988	CBS/New York Times	Helping Middle Class	25%	
		Environment	10%	Dukakis 66% Bush 34%
1990	Voter Research & Surveys (VRS)	Education	26%	
		Environment	21%	
1992	Voter News Service (VNS)	Economy/Jobs	42%	
		Environment	5%	Clinton 72% Bush 14%
1995*	Gallup	Crime/Drugs	43%	
		Environment	2%	
1996	Voter News Service (VNS)	Economy/Jobs	21%	
		Environment	<2%	

*Exit polls were conducted following the November elections of each year cited. Due to preelection polling conducted in 1994 showing negligible interest in the environment as an election issue, the major exit pollsters did not include the environment as a category in the 1994 exit polls. The results of a Gallup poll conducted in early 1995 reflect the lack of salience of environmental issues.

Sources: Various polling organizations; presented in Ladd and Bowman, *Attitudes Toward the Environment*, table 17, p. 46; for 1995 and 1996, author's compilation of polling data.

ment as their most important issue. Interestingly, those stating that the environment was their primary concern, split more evenly by partisan affiliation than in the presidential elections presented, with 55 percent choosing the Democratic congressional candidate in their district and 44 percent choosing the Republican candidate. By the 1992 presidential election, the environment had faded from prominence; two years later there was so little interest among voters that pollsters dropped the environment as a possible response.[19]

Despite the lack of salience of environmental issues in the 1994 congressional elections, the takeover of both houses of Congress by the Republican Party led to a redirection of the Democratic Party issue agenda. By 1996, the Clinton/Gore campaign had included the environment in its four-pronged issue agenda that came to be known as MMEE or, for *Star Wars* fans, M2E2: Medicare, Medicaid, Education, and the Environment. Despite the campaign focus on the environment, particularly as a response to the antienvironmental policies pursued by the new Republican Congress, the environment, as a campaign issue, failed to register as an important factor in the 1996 elections.[20]

Related to the general patterns of public support for the environment and the relative salience of the environment in the electoral context is the portrayal of environmental issues by the news media in the United States.[21] Anthony Downs, writing more than two decades ago, identified "issue attention cycles" in public perceptions of the environment.[22] Downs argued that the significance of issues relating to the environment will rise and fall as the public's attention is drawn to other, more relevant, policy concerns. To the extent that news organizations are driven by "newsworthiness," the allocation of column inches in newspapers or minutes on network news programs is likely to reflect the issue attention cycles of newspaper editors and news producers. Media coverage of the environment, then, will serve as an additional resource for assessing the broad political context within which environmental organizations operate.

The patterns of news reporting on the environment, presented in figure 2.1, clearly demonstrate the Downsian issue attention cycles relating to the environment as manifested in media coverage.[23] There are a number of peaks and valleys in environmental coverage that, in turn, broadly shape what citizens think about in their daily lives. Media scholars have consistently argued that the media play an integral role in setting the policy agendas for the general public.[24] The pattern of coverage by the *New York Times* reflects the broader political salience of environmental issues. Each of the peaks in press coverage captures the media attentiveness to environ-

FIGURE 2.1 **Media Coverage of the Environment**
             Source: New York Times Index

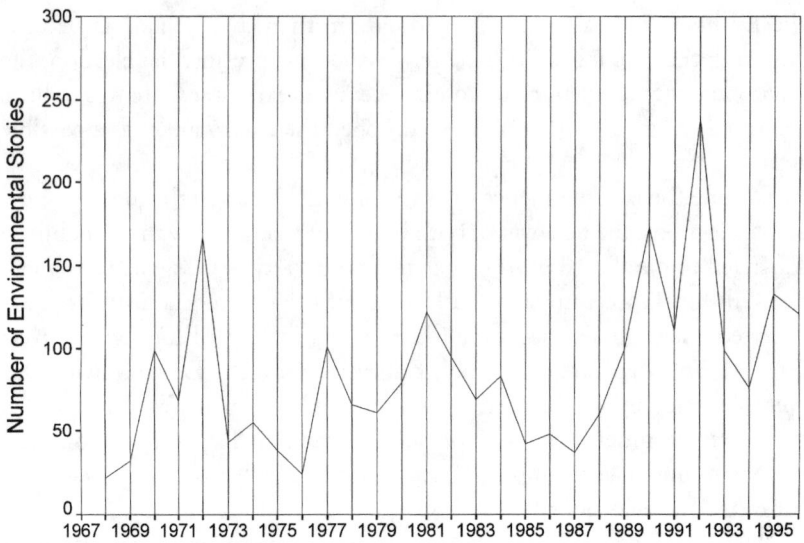

mental issues in particular years. While the sheer number of stories alone does not shed full light on the nature of news coverage of the environment, the volume of stories does indicate the overall importance of environmental issues vis-à-vis other policy issues competing for coverage by the news media.

The 1970 peak in coverage captures the media attention given to the first Earth Day celebration, the creation of the Environmental Protection Agency (EPA) following the passage of the National Environmental Policy Act a year earlier, as well as the passage of the Occupational Safety and Health Act, the Water Quality Control Act, and amendments to the Clean Air Act. The peak in 1972 resulted from coverage of the passage of several significant legislative initiatives, including the Water Pollution Control Act, the Marine Mammal Protection Act, the Pesticide Control Act, and the Coastal Zone Management Act.

Following four years of relative inattentiveness by the media, the early legislative activities following Jimmy Carter's election, including the amendments to the Clean Air, Clean Water, and Ocean Dumping acts as well as the initial proposal of the Carter energy package, resulted in a fourfold increase in environmental news stories from 1976 to 1977. Despite the coverage of the Love Canal contamination in New York and the Three

Mile Island radioactive leakage, news coverage dropped in 1978 and remained fairly level for the next two years.

With the election of Ronald Reagan in 1980 and the subsequent articulation of his environmental agenda, much of the news coverage in 1981 focused on the controversies surrounding his administration's appointees, particularly Watt and Gorsuch Burford mentioned earlier. In subsequent years, media coverage of environmental issues declined precipitously through the first term of the Reagan administration and remained low during his second term in office.

Following the Reagan era, environmental news coverage in 1989 was dominated by the Exxon *Valdez* oil spill in Alaska, but also included analyses of the Bush administration's environmental policies and administrative appointments, including the selection of William Reilly as EPA administrator. The following year, 1990, generated the highest number of news stories up to that point, with coverage of the twentieth anniversary of Earth Day, but also stories on the significant amendments to the Clean Air Act.[25] The defeat of the "Big Green" ballot proposition in California also attracted media attention. Recall from the data in table 2.4 that the environment was identified by more than 20 percent of voters as their most important concern in the 1990 elections, supporting the agenda-setting role of the media. The Persian Gulf War in 1991 relegated much of the environmental policy coverage to secondary status, although the legislative stalemate on the Arctic National Wildlife Refuge Bill received a fair amount of coverage.

In almost three decades of media coverage, 1992 represents the high-water mark for volume of reporting on environmental issues. A number of issues and events factored into the publication of 239 environmental stories by the *New York Times*. Clearly the selection of Albert Gore as the Democratic vice presidential candidate raised the level of attentiveness to environmental concerns, leading to a dialogue on the environment between the Bush and Clinton campaigns. Nonetheless, the bulk of the news coverage in 1992 was not focused on the American political arena, but rather on the "Earth Summit" in Rio de Janeiro, Brazil. The dramatic appearance of global environmental issues in the American media is important for two reasons. First, it raises questions regarding the significance of domestic environmental issues on which the national environmental organizations were largely focused at the time. Second, the exit poll results reported in table 2.4 for the 1992 and 1996 presidential elections, demonstrating that environmental concerns were not prominent in the voting decisions of most voters, raise additional questions regarding the importance of international environmental issues to the American public. In chapter 5, both of these questions will be analyzed in the broader context of membership re-cruitment and retention within the segment of the public attentive to envi-ronmental issues.

In the two years following 1992, environmental coverage not only de-clined but shifted in focus. Increasingly prevalent were stories about the environmental movement itself, about the possibility of decline in the movement's strength, and also about the rise of antienvironmental causes in the form of the "wise-use" movement and the property rights move-ment. In the aftermath of the "Republican Revolution" in the 1994 con-gressional elections, environmental stories took on a defensive character. Interestingly, the 1995 peak mirrors the 1981 increase in media coverage fol-lowing the Reagan election. News stories focused on the environmental consequences of the "Contract with America," in particular on the regula-tory reform agenda of the Republican leadership in the House of Repre-sentatives.[26] By election year 1996, coverage of the environment as a politi-cal or policy issue declined somewhat from the 1995 level of reporting.

Together, these various sources of environmental information regarding the general public and the media serve to define American environmental-ism within the broad political spectrum of issues and ideas. Based on pub-lic opinion responses, the citizenry appears clearly supportive of a clean and healthy environment. But according to Ladd and Bowman, "we are now more inclined to think that for most Americans, the urgency has been removed, and the battle to protect the environment is being waged satis-factorily."[27]

In the electoral context, environmental concerns rarely mobilize signifi-cant numbers of voters. Further, in the event that the issue of the environ-ment is salient, as in 1990, voters are not monolithically supportive of the Democratic Party, as is the case when much smaller blocs of voters are con-cerned with the environment. Finally, media coverage of the environment waxes and wanes, largely driven by the policy agenda of the federal govern-ment in the nation's capital. To the extent that the mass media serve as agenda setters, reliance on the national news media for consistent informa-tion about the environment is suspect. It is within this broad political con-text that the analysis of a distinct attentive public in American society—leaders and members of national environmental organizations—will take place.

The Contemporary Environmental Movement

Within these broad political parameters, the environmental movement op-erates at the national level as more than 100 professionalized, bureaucratized public interest organizations, at times allied in political advocacy, but often in competition for resources and members. The national environmental organizations presented in table 2.5 are representative of groups active in the federal policy-making process today and offer evidence of the organizational diversity in this particular public interest arena. From the more conservative conservation-oriented organizations to the most radical "monkey wrenchers," the contemporary environmental movement includes all shades of green politics as well as diverse organizational structures and strategies of political influence.[28]

The contemporary environmental movement is amply represented by traditional 501(c)(3) organizations, but also includes several full-service environmental conglomerates such as the Sierra Club. The movement includes organizations focusing primarily on litigation (e.g., National Resources Defense Council), on research-driven advocacy (e.g., Environmental Defense Fund), on land acquisition (e.g., the Nature Conservancy) as well as on direct political influence in the electoral process (e.g., League of Conservation Voters). In addition to the majority of advocacy organizations using traditional means of influence (lobbying, research, technical assistance, federal agency monitoring, congressional testimony, coalition formation, and grassroots efforts), there are also several organizations that go beyond social reform tactics by mobilizing for direct action against governments and private economic interests perceived to be in violation of the groups' environmental norms (e.g., Sea Shepherd Conservation Society and Earth First!). Finally, there is Greenpeace, U.S.A., an organizational hybrid that incorporates organizational attributes of social reform organizations with the tactics of direct action groups.[29]

Organizations in the environmental movement, compared to the groups in the broad public interest sector presented in chapter 1, tend to be older, wealthier, less reliant on foundation support, and have larger memberships and staffs.[30] As such, national environmental organizations are representative of more developed public interest organizations in the United States. Leadership of national environmental organizations entails managing these public interest enterprises with budgets ranging from under $1 million to more than $300 million as well as sustaining representative policy advocacy through organizational strategies and interaction with memberships ranging from less than 50,000 to more than 1 million.

Given the organizational diversity within the contemporary environmental movement, it is necessary to focus more narrowly on a representative sample of five organizations that capture the internal diversity in substantive policy agendas, organizational structures, leadership styles, membership size, organizational wealth, and longevity. The five national

TABLE 2.5 National Environmental Organizations

Group	FOUNDING Date	TAX Status	BUDGETS in Millions	Membership	Staff	MEMBERSHIP Support
Center for Marine Conservation	1972	501(c)(3)	$ 8.1	120,000	59	64%
Clean Water Action	1971	501(c)(4)	$13.0	600,000	50	98%
Defenders of Wildlife	1947	501(c)(3)	$ 6.7	125,000	60	45%
Earth First!	1980	501(c)(4)	$ 0.1	10,000	5	100%
Environmental Action, Inc.[a]	1970	501(c)(3)	$ 0.6	12,000	16	42%
Environmental Defense Fund	1967	501(c)(3)	$25.7	300,000	166	59%
Friends of the Earth[b]	1990	501(c)(3)	$ 2.5	20,000	22	14%
The Fund for Animals	1967	501(c)(3)	$ 2.8	100,000	36	25%
Greenpeace USA	1971	501(c)(3)	$37.8	1,600,000	300	70%
Izaak Walton League of America	1922	501(c)(3)	$ 2.3	45,000	24	27%
League of Conservation Voters	1970	501(c)(4)	$ 2.5	30,000	22	100%
National Audubon Society	1905	501(c)(3)	$43.1	550,000	270	22%
National Parks and Conserv. Assoc.	1919	501(c)(3)	$12.3	350,000	50	79%

National Wildlife Federation	1936	501(c)(3)	$97.1	4,000,000	474	44%
Natural Resources Defense Council	1970	501(c)(3)	$26.2	170,000	168	61%
The Nature Conservancy	1951	501(c)(3)	$323.5	828,000	1,804	26%
Sea Shepherd Conservation Society	1977	501(c)(3)	$ 0.5	30,000	1	92%
Sierra Club	1892	501(c)(4)	$43.0	550,000	294	37%
The Wilderness Society	1935	501(c)(3)	$15.2	275,000	123	49%
World Wildlife Fund[c]	1948	501(c)(3)	$79.0	1,200,000	302	37%

[a] Prior to 1994, Environmental Action, Inc., a 501(c)(4) organization, was affiliated with Environmental Action Foundation, a 501(c)(3) organization. In July of 1994, the two entities merged; the new organization operated with 501(c)(3) tax status. In October 1996 Environmental Action ceased operations.

[b] In 1990 Friends of the Earth merged with the Environmental Policy Institute and the Oceanic Society.

[c] In 1990 the Conservation Foundation merged with the World Wildlife Fund.

Sources: Paul McClure, ed., *Public Interest Profiles, 1996–1997* (Washington, D.C.: Congressional Quarterly Press, 1996); Everett Carll Ladd and Karlyn H. Bowman, *Attitudes Toward the Environment: Twenty–Five Years After Earth Day* (Washington, D.C.: AEI Press, 1995), p. 59.

environmental organizations selected for further analysis are the Sierra Club, National Wildlife Federation, The Wilderness Society, Environmental Defense Fund, and Environmental Action. These groups include the oldest of the contemporary environmental organizations as well as two of the newer groups formed in the new environmental era of the late 1960s. The sample includes one of the largest and wealthiest organizations in the environmental movement as well as one of the smallest and least wealthy throughout its history (see tables 2.6 and 2.7). The sample also reflects the organizational volatility of the national environmental movement as one of the five organizations, Environmental Action, was forced to cease operations in October 1996.

Substantively and ideologically these five groups provide disparate policy agendas, both in the breadth of issues they cover and in their philosophical approaches to advocating their policy positions. Their internal decision-making processes are also quite different. At the same time, these

TABLE 2.6 **Organizational Memberships (x 1,000)**

ORGANIZATION	1965	1970	1975	1980	1985	1990	1996
Sierra Club (1892)	27	113	147	182	363	600[a]	550
The Wilderness Society (1935)	20	54	64	45	147	350[a]	275
National Wildlife Federation (1936)	240	540	612	818	900	997[a]	650
		[b](3.1)	(3.7)	(4.0)	(4.5)	(5.8)	(4.0)
Environmental Defense Fund (1967)	—	11	37	45	50	150	300[a]
Environmental Action (1970)	—	10	11	20[a]	15	23	12[c]

[a]Sierra Club membership peaked in 1991 at 650,000 members, but by 1993 membership had fallen to 535,000. The memberships if the National Wildlife Federation and The Wilderness Society peaked in 1990. Environmental Defense Fund membership continues to grow as 1996 membership surpassed 300,000. Environmental Action membership peaked in the early 1980s at 26,000.

[b]The National Wildlife Federation also includes as members those individuals belonging to their federated state and local chapters. Those membership figures (in millions) are presented in parentheses.

[c]Environmental Action ceased operations in October 1996 with approximately 12,000 members.

Sources: "National Environmental Organization Memberships, 1960–1983," personal correspondence with Robert C. Mitchell, Resources for the Future; Stephen Fox, *The American Conservation Movement* (Madison: University of Wisconsin Press, 1985), p. 314; Paul McClure, ed., *Public Interest Profiles, 1996–1997* (Washington, D.C.: Congressional Quarterly Press, 1996); interviews with membership and development directors.

TABLE 2.7 Organizational Budgets (in $ millions)

ORGANIZATION	1981	1985	1987	1990	1995
Sierra Club	12.0	21.4	23.0	35.0	43.0
The Wilderness Society	3.0	6.3	9.4	17.9	15.2
National Wildlife Federation	32.0	46.0	59.3	87.2	97.1
Environmental Defense Fund	2.6	3.4	6.1	15.1	25.7
Environmental Action	0.6	0.5	0.3	1.1	0.6

Sources: Public Interest Profiles, (1982, 1986, 1988, 1992, 1996); interviews with membership and development directors.

groups have organizational characteristics in common. Each of these organizations recruits and maintains its members in a similar fashion. Direct mail serves as the primary method of membership recruitment and retention for national environmental organizations, with telemarketing efforts used as supplemental tools, particularly for the retention of existing members. Each organization has an organizational structure that includes a board of trustees and the executive leadership of the organization, although the relationships between these are quite dissimilar.

Finally, the focus of the political representational effort is similar for each of these organizations. The federal environmental policy-making infrastructure, including the United States Congress, the White House and the relevant executive branch departments and agencies, and the federal courts, comprises the primary arena in which the political representation of environmental interests takes place.

Before examining these organizational differences and similarities in the context of organizational maintenance and political representation in the contemporary environmental movement, a substantive overview of each of these organizations may be helpful in understanding the analyses of organizational leadership, leadership-membership relationships, and the consequences of these organizational attributes for coherent political representation in the chapters that follow.[31]

National Environmental Organizations: Five Case Studies

Sierra Club: "One Earth, One Chance"

The Sierra Club,[32] founded in 1892, is the oldest and most organizationally complex of the contemporary environmental organizations. The parent organization, Sierra Club, Inc., is a nonprofit, 501(c)(4) public interest organization that pursues a broad conservation agenda and attempts to influence the policy process not only through advocacy in the legislative, executive, and judicial arenas but also indirectly through the electoral process. Its mission is "to explore, enjoy, and protect the wild places of the earth; to practice and promote the responsible use of the earth's ecosystems and resources; to educate and enlist humanity to protect and restore the quality of the natural and human environment; and to use all lawful means to carry out these objectives."

Organizationally, the Sierra Club is what I have identified in my earlier work as a "full-service" public interest organization.[33] Rather than having all organizational activities centralized under one nonprofit entity, the Sierra Club infrastructure is organized, for tax purposes, as a series of related yet distinct entities. In addition to the parent 501(c)(4) organization, the Sierra Club network includes the 501(c)(3) Sierra Club Legal Defense Fund (renamed the Earth Justice Legal Defense Fund in 1998), the 501(c)(3) Sierra Club Foundation, and Sierra Club PAC, a political action committee. Also operating within the Sierra Club infrastructure are the Sierra Student Coalition, a student activist program founded in 1991, Sierra Club Books, the publishing wing of the organization, and Sierra Club Property Management, Inc., a wholly owned subsidiary charged with managing its headquarters property.

The full-service model of operation is useful for several reasons. First, it allows the parent organization to conduct its direct policy advocacy activities, unencumbered by the IRS limitations placed on 501(c)(3) nonprofit organizations.[34] Second, the parent organization may receive funding from its related 501(c)(3) entities. During the past five years, Sierra Club, Inc., has received grants of $3–4 million annually from its 501(c)(3) Foundation. Under IRS rules, such grants are allowable as long as the funds are not used for direct lobbying purposes. This relationship allows the Sierra Club to direct its large donors to the foundation where such contributions are tax deductible. Third, 501(c)(4) organizations are allowed to have affiliated political action committees; 501(c)(3) groups may not be involved in electioneering activities.

In the 1994 congressional elections, Sierra Club PAC ranked as the tenth largest ideological, single-issue PAC with contributions to candidates of more than $400,000.[35] By the 1995–1996 election cycle, the Sierra Club was fully mobilized for political and electoral battle as it far surpassed its previous efforts. In these two years, the Sierra Club and its PAC spent $7.5 million in issue advocacy campaigns, voter education drives, and direct electoral activities.[36]

Beyond its national headquarters in San Francisco, the Sierra Club has a Washington, D.C., office, directed by Daniel Weiss, that serves as the lobbying wing of the organization. In addition, the Sierra Club is organized through twenty-two regional offices as well as state and local chapters and provides for its members a variety of activities through which members are able to interact with each other. This organizational approach, while providing one of the few examples of the direct linkage of members and leaders in the contemporary national environmental movement, is quite costly. For the third year in a row, the Sierra Club began the 1995 fiscal year with a budget deficit of more than $3.0 million. In an effort to control costs, the Club embarked on a painful process of reducing its professional staff by 10 percent by the end of 1995. The organizational strife was eased a bit by a membership increase during the year of almost 20,000 new members. Due to its complex structure with multiple sources of revenue production, the Sierra Club is less reliant on membership dues than most national environmental organizations, with just over one-third of its income derived from annual dues (see table 2.6 for comparisons).

The renewed membership outreach efforts of the Sierra Club are reflective of the organizational ethos that has guided the Club for more than 100 years. Today the Sierra Club attracts more than half a million members for a wide variety of reasons. Some people join the Club for its purposive goals of preserving wild lands and wildlife through lobbying and grassroots efforts. Others may enjoy the wide selection of magazines and newsletters that the Club provides for its members, from its flagship *Sierra* magazine (with a paid circulation of 511,000 in 1995),[37] to its activist newsletter, *The Planet,* or SC-ACTION mailings that contain action alerts, news releases, or other items of national interest, or *The Student Activist* for college students. These members may also enjoy the publications of the regional, state, and local chapters. Still other members may enjoy the discounts they receive on Sierra Club books and merchandise or the worldwide outings programs offered by the Club. More than likely Sierra Club members are attracted to the organization for some combination of these incentives. For a base membership fee of $35.00 ($15.00 for students), Sierra Club members receive a variety of benefits.

The Wilderness Society

Founded in 1935, The Wilderness Society celebrated its sixtieth anniversary amid significant organizational restructuring. Operating as a 501(c)(3) nonprofit organization, The Wilderness Society, a group dedicated to the protection of America's wildlands and wildlife, is "the only national conservation organization to focus on issues affecting national parks, forests, wildlife refuges, and Bureau of Land Management lands." The organization attempts to foster its mission of a land ethic through a variety of efforts including public education, advocacy, and economic and ecological analysis.

In addition to promoting a land ethic, The Wilderness Society seeks to protect wildlands, parks, forests, rivers, deserts, and shorelands across the United States, including the ancient forests of the Pacific Northwest and northern California, the Arctic National Wildlife Refuge in Alaska, the Colorado Plateau area, and the Southern Appalachian forests, as well as forests in New England and upstate New York. The Society has as one of its broad goals the doubling of the size of the National Wilderness System by 2015; it also seeks to reform several public lands subsidy programs of the federal government, including logging, mining, and grazing programs. Protection of endangered species is also an important component in the Society's policy agenda.

In 1997 its membership included more than 275,000 citizens nationwide and provided roughly half of the funding for the Society. Its operating budget of approximately $15 million reflects a slight rebound from the early 1990s but does not match its largest budget attained in 1990. As a result of the decline in budgetary support in the early 1990s, the Society staff dipped to a low of ninety-seven employees. It now operates from its national headquarters in Washington, D.C., as well as from nine field offices across the country in Seattle, San Francisco, Boise, Bozeman, Denver, Minneapolis, Boston, and Atlanta. Prior to staff layoffs, its 123 employees—just short of its largest staff of 125 in 1992—are located in Washington, D.C., and fourteen regional offices.

The financial burdens of the early 1990s have also altered the communications processes within the organization. Throughout the contemporary environmental era, The Wilderness Society has published a quarterly magazine, first called *Living Wilderness* and now simply *Wilderness* (with a paid circulation in 1995 of 195,000 copies). In April 1996 the Society reconstituted its publications by offering *Wilderness* only twice a year in addition to a quarterly newsletter, *Wilderness Watch*. The organization also offers *Wild Fish and Forests* (dealing with wildlife issues in the Northwest) and *New Voices* (an activist newsletter) to smaller constituencies within its membership. The Wilderness Society tends to attract members who are specifically focused on wilderness conservation issues rather than the more contemporary environmental issue agenda that includes toxic waste management, pollution, nuclear power, pesticides, and global environmentalism.

Unlike the Sierra Club, The Wilderness Society offers few programs or services that enable members and leaders to interact in a direct manner. The regional offices throughout the country pursue the substantive policy issues of the Society rather than provide constituency services for members.

National Wildlife Federation

The National Wildlife Federation was founded in 1936, one year after The Wilderness Society, as a result of the First North American Wildlife Conference convened by President Franklin D. Roosevelt. "The mission of the National Wildlife Federation is to educate, inspire and assist individuals and organizations of diverse cultures to conserve wildlife and other natural resources and to protect the Earth's environment in order to achieve a peaceful, equitable and sustainable future." The group's main goal is to raise awareness and involve people of all ages in their fight to conserve and protect the environment by bringing together interested organizations, governmental agencies, and individuals on behalf of restoration of land, water, forests, and wildlife resources.

Self-described as the nation's largest, nonprofit conservation education organization, the National Wildlife Federation often boasts of its more than four million members and supporters, as reported in table 2.6. The Federation membership figures presented in table 2.6, however, distinguish between members and supporters. In a comparative context, the National Wildlife Federation has 650,000 members who pay annual dues to the national organization in much the same manner as members of Sierra Club and The Wilderness Society. In addition to the 650,000 "associates," more than four million people are members of the 51 affiliated state organizations or are members of local chapters of clubs with ties to the Federation.

Organizationally, the National Wildlife Federation is a large bureaucratic entity, although not as structurally diverse as the Sierra Club. There are two main structures that comprise the national organization: the National Wildlife Federation, and the National Wildlife Federation Endowment, Inc.; both are 501(c)(3) nonprofit, tax-exempt organizations. The bulk of the Federation activities flows from the parent organization, while the endowment serves as the repository of large financial donations as well as the financial investment branch of the Federation. There are two additional entities within the Federation infrastructure that operate as subsidiaries: National Wildlife Productions, Inc., a film and television video production firm, and International Wildlife Incorporated, "a nonprofit entity whose initial objective is to contribute to wildlife appreciation through the retail gallery sale of art works created by internationally acclaimed painters."

Recently the Federation sold off its interest in its former national headquarters building in Washington, D.C., and moved its operations to its facility in the Washington suburbs of northern Virginia. The Federation also maintains nine regional resource centers across the country. Operationally, in addition to its senior leadership of nineteen executive officers, the Federation receives input from thirteen regional directors. The national headquarters organization includes six departments: resources conservation, the lobbying and advocacy division, international affairs, educational outreach, affiliate and regional programs, public affairs, and the campus outreach program (Campus Ecology Project), created in the late 1980s to promote environmental awareness on college campuses.[38]

Operating today with a budget of just under $100 million and a staff of almost 500, the National Wildlife Federation functions as an environmental conglomerate with complex educational, publishing, research, and advocacy operations.[39] The educational and publishing divisions of the Federation produce the bulk of the annual revenues. While the Federation reported that 44 percent of its budget was derived from membership dues in 1994 ($44 million), only 12 percent was generated from associate and life memberships. Of the remaining 32 percent, roughly half of these revenues were generated through donations and bequests ($14.6 million) and the other half through the sale of the Federation's popular children's magazines, *Ranger Rick* and *Your Big Backyard* ($16.8 million).[40]

In addition to the magazine sales, the Federation earned more than $44 million from the sales of nature education materials. Including income from outings programs and royalty income on publications, fully two-thirds ($65.6 million) of the National Wildlife Federation revenues were produced through the sales of organizational goods and services.[41] Through its programs and magazine sales as well as its broad issue advocacy efforts, the National Wildlife Federation attracts a large and diverse membership.

Environmental Defense Fund (EDF)

After the creation of the Sierra Club and other old-line conservation groups in the late nineteenth century and the second wave of conservation organizations such as The Wilderness Society and the National Wildlife Federation, founded in the interwar period, came a new wave of environmental organizations in the late 1960s and early 1970s. The first of the third wave groups to be formed was the Environmental Defense Fund.

As stated earlier, EDF was founded by a group of scientists alarmed by the effects of DDT on the wildlife on Long Island. Joining with an attorney, the scientists became the first litigants to use scientific evidence to achieve environmental goals through the federal court system. With the support of the Ford Foundation as well as newly allied organizations such as the National Audubon Society, through its Rachel Carson Memorial Fund, EDF was well on its way to formulating numerous cases to attack the use of DDT as a means of pest control. Within five years, DDT was banned by the newly created U.S. Environmental Protection Agency.

This initial success strengthened the fledgling organization but also guided other groups to seek remedies to environmental problems in the courts, including the Sierra Club, through the creation of its Legal Defense Fund, the National Wildlife Federation, and the National Audubon Society. The successes also lead to the creation of new environmental litigation firms, such as the Natural Resources Defense Council, founded in 1970 with support from the Ford Foundation.

Having celebrated its thirtieth anniversary in 1997, the Environmental Defense Fund is fiscally sound. The organization began 1997 with more than 300,000 members, the largest membership base in its history. Its 1996 operating budget of $25.7 million was also the highest in EDF's history, with roughly 60 percent of the budget generated from membership dues and contributions. The current staff of 166 employees, including more than 60 full-time scientists, economists, and attorneys is also the largest the organization has ever had.

Headquartered in New York City, EDF has five regional offices located in Washington, D.C., Oakland, Boulder, Raleigh, and Austin; it also operates a project office in Boston. EDF operates as a 501(c)(3) nonprofit organization and provides few of the goods and services offered by the National Wildlife Federation or the Sierra Club. Members receive an eight-page, bimonthly newsletter for their annual dues of $35.00.

Substantively, EDF focuses on a broad range of regional, national, and international environmental issues that include the organization's *9 Issues for the 90's:* greenhouse effect, wildlife habitat, ozone depletion, saving the rainforests, acid rain, Antarctica, toxics, water, and recycling. The broad theme of market-based environmental problem solving pursued by the Environmental Defense Fund during the 1990s is reflective of the survey findings presented earlier in this chapter.

"EDF believes a healthy economy and a healthful environment are not mutually exclusive—they can go hand in hand. But we must make sure the rules of the game don't reward actions that pollute the environment or endanger human health." This message offers a distinctive counterpoint to much of the doomsaying put forth by a number of national environmental organizations. The message clearly has resonance for a growing number of environmental activists as EDF doubled its membership from 1990 to 1995, while many national organizations struggled to maintain existing membership levels.

Environmental Action (Foundation)

In this analysis, Environmental Action serves to represent the small wing of the national environmental movement that has sought to retain the organizational ethos present in the movement at its inception more than twenty-five years ago. Ultimately the efforts of Environmental Action failed to keep pace with the organizational realities of the late 1990s. Nonetheless, its quarter-century of environmental advocacy serves as reminder of what once was the norm in the environmental movement.

Founded in 1970 by the organizers of the first Earth Day, Environmental Action struggled through its twenty-six years of operation as one of the smallest, yet historically one of the most tenacious of the national environmental organizations. Prior to July 1994 Environmental Action operated as two nonprofit entities—Environmental Action, Inc., a 501(c)(4) organization, and Environmental Action Foundation, a 501(c)(3) organization. Affiliated with the 501(c)(4) entity was EnAct/PAC, a small political action committee known more for the publication of its "Dirty Dozen" list of congressional incumbents with the worst environmental voting records than for its financial contributions to candidates.

Between 1994 and 1996, the organization operated as a 501(c)(3) nonprofit under the name of Environmental Action Foundation, but remained commonly known as Environmental Action. Even up to the very end, the leadership and staff sought to maintain the organization through downsizing: "this streamlining of our administrative operations will enable us to reach out to a larger audience and strengthen the impact of our work." The unfortunate consequence of the restructuring was the loss of the tenacity that had been the hallmark of the organization. Recall that 501(c)(3) organizations may not engage in electioneering through direct support or opposition, either financially through a PAC or through endorsements or targeted opposition, such as the "Dirty Dozen" hit list. In 1995, Environmental Action turned over its "Dirty Dozen" campaign to the League of Conservation Voters.

At the end, with a staff of sixteen, Environmental Action operated entirely from its new headquarters in Takoma Park, Maryland, a Washington, D.C., suburb, having left its earlier home, a weathered brownstone building off of Dupont Circle in Northwest D.C., as it was beginning to consider restructuring in 1993. The organization was equally reliant on the dues and contributions of its less than 15,000 members and small foundation grants, with approximately 75 percent of its budget of under $1 million derived from these sources in 1995. For their $25.00 annual dues ($15.00 for students and seniors), members received *Environmental Action* magazine, a very substantive quarterly publication.

The Environmental Action policy agenda was largely guided by an overarching vision—"a clean, unpolluted environment is a fundamental human right." Its agenda was focused on environmental rather than conservation issues; it included a series of campaigns targeting industrial pollution concerns such as solid waste and toxics, alternative energy sources as well as energy conservation, environmental justice for low-income and minority communities, and maintaining and bolstering existing environmental regulations at all levels of government. Environmental Action staffers worked with grassroots organizations and a variety of coalitions in Washington as well as in state capitals across the country to promote their policy goals until the organization's end in late 1996.

Having established the broad political context within which contemporary national environmental organizations function and having outlined the organizational characteristics of the representative sample of five environmental groups, the internal operations of national environmental organizations, with a specific focus on these five environmental organizations, will be discussed next. The analysis in chapter 3 provides a broad overview of the executive leadership in the contemporary environmental movement. Focusing on leadership turnover, salaries, responsibilities, division of labor, and the allocation of resources toward membership recruitment and retention, that chapter chronicles the growing pains associated with the rapid expansion of the 1980s followed by the more difficult period of maintenance in the early 1990s. The five environmental organizations are analyzed to assess the leadership structures and management styles that guided each group during the past two decades.

3

Growing Pains
Leadership Challenges in Contemporary Environmental Organizations

A program without an organization is a hoax.
—John Gardner, founder, Common Cause

There's no need to be sloppy and inefficient just because we're a liberal cause.
—William Turnage, past president, The Wilderness Society

Nonprofit leadership in general and public interest leadership in particular is no easy task. The job of operating a multimillion dollar enterprise, whether it is he United Way, a public interest organization, or a university for that matter, is multifaceted and, at times, perilous. Leaders of nonprofit organizations serve a variety of constituencies simultaneously—members (donors or students), paid and volunteer staff, boards of directors, and major financial supporters, among others. They also must bring to their jobs a variety of leadership capabilities—management skills, entrepreneurial skills, interpersonal skills—and a political vision that will direct the organization.

The difficulty in matching leaders with organizations lies in understanding what combination of attributes are necessary to lead a particular organization at a particular time. At a crucial point in an organization's existence, it may need a leader who is able to transform the core mission of the organization. At other times, the organizational mission may be quite clear, but the organization desperately needs to be managed more effectively. These two scenarios call for different types of leaders. For many contemporary nonprofit organizations, deciding on the requisite leadership needs at a particular time is the most difficult task they face.

A colleague of mine at a neighboring public university succinctly summarized the outcome of what he believed to be the inappropriate hiring of a new president: "We needed a cowboy and we got a manager." What he meant by this statement is that at this particular time his university was functioning well on paper but lacked a fresh vision of its future. He was hoping for a charismatic, transformational leader to guide the university into the next century. Instead, the university hired, in his estimation, a bureaucratic administrator with no demonstrable skills to redirect its core mission. Such problems in leadership pervade the public interest sector, and the national environmental movement in particular, for a variety of reasons. First, the internal operations of these groups include a wide variety of organizational structures and processes, making leadership decisions more complex. Second, of the many arenas within the public interest sector none has the varied history of the environmental movement, with its significant changes in organizational structures and in substantive policy agendas. Third, the transformation from social movement groups to public interest organizations and the growing pains associated with the transformation are evident in the various organizations comprising the contemporary environmental movement.

Historian Stephen Fox has argued that the American conservation movement emerged early in this century as a "hobby" or "avocation," but over time, through institutionalization and an infusion of science into the art of conservation, has transformed itself into a profession.[1] A former leader of one of the largest and most prominent conservation organizations in this country agrees with this argument but acknowledges the consequences of such professionalization:

> There are so many more organizations today than when I began. There are many more outlets and niches for leadership. The increases created more diversity in the community and more opportunity for entrepreneurialism and experimentation. I think that's very healthy. But the quality of leadership? Well, you get into the Stephen Fox thesis, that the earlier leaders came out of a volunteer tradition and tended to be stronger, more assertive. They were outspoken, risk takers. On the flip side of the coin, they were less good at being managers. Many weren't managers at all.
>
> I suppose Fox's supposition is true, but there are people in recent years who feel strongly and care deeply, too. Trouble is, there is this crisis as organizations grow. They keep outgrowing themselves—their leaders are not good enough as managers.[2]

These tensions between organizational management skills and substantive expertise on the part of environmental leaders became even more acute as organizations grew in size and wealth through the end of the 1980s. While the policies of the Reagan administration served as substantive catalysts for growth in the organized environmental sector in the 1980s, additional organizational changes within groups facilitated the rapid expansion and institutionalization, including the increased use of direct mail and other mass-marketing methods. Regardless of the sources, the rapid expansion and resulting restructuring of the movement into a complex, multibillion dollar public interest enterprise was undertaken in relatively short order, leading to significant organizational growing pains.[3]

Evidence of the organizational change that resulted from the growing pains of the last fifteen years is manifest in the executive leadership turnover within many national environmental organizations. Of the five organizations identified in the last chapter, only the Environmental Defense Fund is still directed by the same person as in 1985 when I began this study, Fred Krupp; nevertheless there were two leadership changes in the EDF in the early 1980s. While long-term leadership is not unique in the environmental movement, the pattern of leadership turnover within many national organizations is quite different.

Since 1980 the Sierra Club has had four executive directors, with the current leader, Carl Pope, appointed in 1992. Similarly, The Wilderness Society has had six leaders (presidents) during the same time period, including the selection by the Governing Council of William H. Meadows III in September 1996. The National Wildlife Federation has also experienced a recent leadership turnover when its longtime president and chief executive officer, Jay Hair, was replaced with Bill Howard in 1995, who was then succeeded by current president Mark Van Putten in June 1996. Environmental Action, like the Sierra Club, also had four executive directors in the last fifteen years (see table 3.1 for comparisons with other environmental organizations).

While some of the leadership transitions that have occurred within environmental organizations in the past two decades were not the result of significant internal strife, most of the leadership changes in the past five years are directly related to the rapid organizational growth during the 1980s and to the subsequent membership decline and budgetary stagnation of the early 1990s. Before assessing the circumstances leading to the various changes in organizational leadership within these national environmental organizations, it is necessary to analyze the broader meaning of leadership within the public interest sector.

TABLE 3.1 Environmental Leadership, 1980–1998

	EXECUTIVE DIRECTORS/PRESIDENTS
Defenders of Wildlife	Allen E. Smith
	William G. Painter (1985–1986)
	Joyce M. Kelly (1986–1987)
	M. Rupert Cutler (1987–1991)
	Rodger Schlickeisen (1991–present)
Environmental Action	Elizabeth A. Davenport
	Alden Meyer (1982–1985)
	Ruth Caplan (1985–1993)
	Margaret Morgan-Hubbard (1993–1996)
Environmental Defense Fund	Janet W. Brown
	William V. Brown (acting: 1983)
	Frederic D. Krupp (1984–present)
Izaak Walton League	Jack Lorenz
	Maitland Sharpe (1991–1995)
	Paul Hansen (1995–present)
League of Conservation Voters	Marion Edey
	Alden Meyer (1985–1987)
	James Maddy (1987–1995)
	Debra J. Callahan (1995–present)
National Audubon Society	Russell Peterson
	Peter A. A. Berle (1985–1995)
	John Flicker (1995–present)
National Wildlife Federation	Thomas Kimball
	Jay D. Hair (1981–1995)
	Bill Howard (1995–1996)
	Mark Van Putten (1996–present)
Natural Resources Defense Council	John H. Adams
Sierra Club	Michael McCloskey
	Douglas Wheeler (1985–1987)
	Michael L. Fischer (1987–1992)
	Carl Pope (1992–present)
The Wilderness Society	William A. Turnage
	George T. Frampton (1986–1993)
	Karin Sheldon (acting: 1993)
	G. Jon Roush (1993–1996)
	Mary F. Hanley (acting: 1996)
	William H. Meadows III (1996–present)

Executive leadership of national public interest organizations involves much more than simply advocating political issues in the federal policy-making process. Nonprofit management skills are necessary to operate these increasingly complex political enterprises. According to the 1995 Foundation for Public Affairs organizational data presented in chapter 1, roughly two-thirds of the sample groups had multimillion dollar budgets and more than ten full-time staffers; one-third of the organizations had more than 100,000 members in 1990 (see tables 1.2, 1.4, and 1.5). Of the national environmental organizations presented in chapter 2, approximately two-thirds have budgets of more than $5 million, memberships greater than 100,000, and staffs of at least fifty employees (see table 2.6). Managing the day-to-day operations of these large national groups requires that executive directors possess a variety of organizational, management, and leadership skills. Today, the pool of candidates with the requisite combination of skills to direct national public interest organizations is not very large.

Beyond the inherent biases in the executive selection process that have resulted in white males being predominant in the executive leadership of nonprofits in general, the biases that exist within the contemporary environmental movement also impede many organizations in their efforts to recruit leaders of high quality. Within the national environmental leadership community, the broader bias in favor of white males in leadership selection identified in chapter 1 is even more pronounced. According to the Foundation for Public Affairs data for 1985, only 7 percent of the environmental organizations in the sample were directed by women; less than 20 percent of the positions on their governing boards were held by women.[4]

As table 3.1 shows, only eight women appear on the list of thirty-seven executive directors of ten environmental organizations from 1980 to the present (22 percent). Three of the eight women have directed the now defunct Environmental Action, the smallest and least wealthy organization in the sample. Two additional women, Karin Sheldon and Mary Hanley of The Wilderness Society, each served less than one year as "acting" presidents at times of leadership turnover. Today, there are only two women, Debra Callahan of the League of Conservation Voters and Kathryn Fuller of the World Wildlife Fund, presiding over widely recognized national environmental organizations.

The national environmental leadership "community" in the United States is a relatively closed, cliquish cadre of political entrepreneurs.[5] Nonetheless, in recent years the method of leadership selection within environmental organizations has often led to the hiring of candidates from outside the individual organizations and even from outside the environmental movement community. Prior to the expansion of the 1980s there was a greater tendency within environmental and conservation organizations to identify leaders internally. As organizations were smaller and less bureaucratized, groups were run, in essence, by a circle of friends. Elevation to leadership occurred through internal consensus. Today, this approach is rarely applied to leadership selections. Of the executive directors listed in table 3.1, less than ten were hired from within the organization. Approximately one-third of the executive directors listed had previously served as leaders of other environmental organizations. The remaining directors were drawn from the larger nonprofit sector, from government service, from the legal profession, or from the private sector business community.

Allen Smith, former director of the Defenders of Wildlife, identifies this increasingly prevalent pattern of hiring from outside the "community culture" as detrimental to the long-term well-being of the movement: "the community isn't growing its top leadership."[6] To some degree the environmental leadership community has acknowledged this shortcoming and has begun instituting management training programs. Additionally, there is a fair amount of formal and informal communication between the leaders and staffs of these organizations that provides a greater sense of cohesiveness. In the early 1980s the major national organizations in the movement organized themselves through the "Group of Ten."[7] By the end of the decade, the larger "Green Group" facilitated coalitional activities within the more broadly defined environmental community.[8] Nevertheless, today's environmental leaders continue to struggle to keep pace with the rapid changes that have transformed the environmental movement. One aspect of executive leadership that reflects the professionalization of the contemporary environmental movement and the public interest sector more generally is the pay scale for executive directors of national organizations. The rise in salary and benefit packages has allowed public interest organizations to be somewhat competitive in attracting executive leadership from disparate fields of the public and private sectors. According to a 1997 *National Journal* study of top executives in the Washington interest group community ($N = 413$), including public interest groups, trade, industry, and professional associations, and unions, nonprofit public interest executive directors receive, on average, a salary of $128,948.[9] In comparison with the larger public interest sector, environmental leaders receive more attractive compensation packages (see table 3.2). Presiding officers of national environmental organizations in the sample received an average salary of $152,218.[10]

TABLE 3.2 Salary and Benefits: Chief Executive Officers

ENVIRONMENTAL ORGANIZATIONS

Center for Marine Conservation	$135,806
Conservation Intnatl. Foundation	172,867
Defenders of Wildlife	140,250
Environmental Defense Fund	254,579
Greenpeace	*65,000[a]*
National Audubon Society	173,417
National Wildlife Federation[b]	336,377
Natural Resources Defense Council	358,216
Nature Conservancy	204,080
Sierra Club	101,049
The Wilderness Society	150,000

OTHER PUBLIC INTEREST ORGANIZATIONS

Alliance for Justice	*74,000[a]*
American Assoc. of Retired Persons	357,201
American Civil Liberties Union	163,525
Children's Defense Fund	*107,759[a]*
Christian Coalition	192,077
Citizens for a Sound Economy	143,530
Common Cause	*86,588[a]*
National Rifle Association	203,519
National Taxpayers Union	58,779
People for the American Way	94,130
Public Citizen	*84,853[a]*

Source: Matthew Reed Baker, "What Interest Groups Pay Their Leaders," *National Journal*, April 26, 1997, pp. 814–815.

[a]Organizations directed by women at the time of the survey: Barbara Dudley, Greenpeace; Nan Aron, Alliance for Justice; Marian Wright Edelman, Children's Defense Fund; Ann McBride, Common Cause; Joan Claybrook, Public Citizen.

[b]Salary and benefit package of Jay Hair, president and CEO of NWF at the time of the survey.

Again, the bias toward males is evident in the salaries accorded men and women across the entire interest group leadership spectrum, including the public interest sector, as well as within the environmental leadership ranks.[11] The salary packages for seventy-two public interest executives are presented in the *National Journal* study. Twenty-one of the seventy-two directors in the analysis were women (29 percent), but one woman and two men reported no financial compensation for their services. When the three uncompensated executive directors are removed from the sample, the remaining sixty-nine executive directors received an average salary of $128,948. The twenty female presiding officers earned an average salary of $108,606; their male counterparts earned an average salary of $137,251.[12] Within the environmental leadership ranks, only Barbara Dudley, then executive director of Greenpeace, appears in the survey. Given that the average compensation package received by environmental directors is more than $150,000, Dudley's salary of $65,000 fell significantly below the average for the movement.

It is within these organizational contexts that contemporary environmental leadership operates. Through the analysis of changing leadership styles of executive directors or presidents of the Sierra Club, The Wilderness Society, the National Wildlife Federation, the Environmental Defense Fund, and Environmental Action, the interactions with their governing boards and professional staffs as well as their memberships the impact of various organizational interactions on the maintenance of national environmental organizations and the ramifications of these organizational relationships in the recruitment and retention of group members can be assessed.

For each of the five organizations, leadership turnover has occurred in the past fifteen years. Such leadership changes often have both positive and negative repercussions. According to leadership expert John Gardner, leadership "renewal" is vital to the longevity and vitality of organizations. Leadership turnover is one important means of renewing public interest organizations.[13] However, James MacGregor Burns argues that changes in leadership may or may not "renew" organizations in the sense meant by Gardner. Burns distinguishes between transactional leadership, or leadership within the existing rules and structures, and transformational leadership, or leadership that truly renews the organization by changing the organizational structures, norms, methods, and perhaps its very reason for being.[14]

For the most part, environmental leadership is transactional in nature; however, on occasion, a transformational leader is introduced into the public interest sector. Gardner himself fits the description through his creation of Common Cause. Ralph Nader also has demonstrated the characteristics of transformational leadership in his public interest career. Whether transactional or transformational, leaders of national environmental organizations continue to face a key challenge—overcoming the organizational inertia that has developed as a result of the dramatic bureaucratization of the organizational infrastructure during the 1980s. This bureaucratization occurred in response to both the shifts in the political environment at the federal and state levels and to the changes in the increasingly competitive political marketplace of fund-raising and membership recruitment. The Sierra Club, perhaps more than any other environmental organization, has faced this challenge with great intensity, both in the acknowledgment of its overexpansion as well as in its response to the organizational challenge.

Sierra Club: Re-Hitched Together

John Muir, the Sierra Club's first president, argued more than a century ago that society must deal with the environment as if everything is "hitched together."[15] Muir's advice is equally applicable to the organization of the Sierra Club today. Despite organizational downsizing and budgetary cutbacks in recent years, the Sierra Club still maintains the most complex and structurally representative organizational decision-making process in the contemporary environmental movement. It is one of the few public interest organizations in which individual members at the grassroots level are able to offer organizational and policy-related input.

The organization is governed by a national board of directors, comprised of fifteen members of the Club, duly elected by the entire membership, or at least by those members wishing to participate in the election process (approximately 20 percent of Sierra Club members vote through mail-in ballots). Officers of the Club include president, largely a symbolic position, vice president, secretary, treasurer, and a fifth officer, elected annually by the board, and an executive director, appointed by the board and accountable to it for the day-to-day operations of the organization. There are also eight issue vice presidents and thirteen regional vice presidents. Today, in addition to the president, Charles McGrady, elected in May 1998,[16] and the executive director, Carl Pope, there exists a position of chairman, currently held by Michael McCloskey, that provides a link between board and staff. Together, Pope, McGrady, and McCloskey provide the executive leadership for the Sierra Club.

Prior to 1994, there also existed the Sierra Club Council below the board and officer levels; it was composed of one delegate from each state or regional chapter, of which there were fifty-nine, including two Canadian chapters and twelve chapters within the state of California. Also represented on the council were delegates from each of the sixty-three committees established by the board of directors. Today, the fifteen members of the national board are referred to as the council. Furthermore, there are no longer sixty-three committees. According to Carl Pope, the Club has made significant inroads toward consolidating and streamlining its operations; "we have, for example, cut 63 committees to six."[17] Beyond the organization at the national level, the Sierra Club is organized at the chapter and group levels as well. Each of the more than fifty chapters has an elected executive committee charged with organization of the membership at the chapter level. At the local level the Sierra Club is organized into more than 300 groups. The vast majority of the Sierra Club chapters are subdivided in this manner.

Sierra Club Chairman Michael McCloskey, who began working at the Sierra Club in 1961, characterized the Club's method of establishing policy priorities through this labyrinthine organizational structure as a "ritualized process."[18] Prior to the budgetary shortfalls beginning in 1991, the agenda-setting process entailed choosing issue priorities for the two-year cycles of each Congress. For each two-year period, six to eight major priorities were established through a process of nomination of issues from the group, chapter, and regional organizations. Governing committees at each level were polled in early fall, prior to the beginning of the first session of Congress, to determine the priorities at the local and state levels. By October the ballots were tabulated, with the issues receiving the most votes getting highest priority. In January, the board of directors met in order to approve the list of issues upon which the Club would focus for the next two years.

As a result of the budgetary shortfalls of the early 1990s leading to the more than $3 million deficit in 1993, the Sierra Club has streamlined its decision-making process and limited the number of issues to be pursued in subsequent Congresses. "It's going to hurt," Carl Pope said; "we might have four top [issues in Congress], and we'll be just as effective with the top three, but not the fourth."[19] Even with the staff cutbacks and the streamlining of the decision-making process, tensions still exist between local and national leaders as well as between national leaders and the professional staff regarding the deliberative nature of decision making; however, the tensions are not as pronounced as they were in the late 1980s when the organization was free of budgetary constraints.[20]

David Brower, who joined the Sierra Club in 1933 and was its first executive director, serving from 1952 to 1969, has derisively compared the decision-making efforts at the Sierra Club to a popular cheese food: "It's becoming like Velveeta; everything must be processed";[21] more generally he has chided the Club, "What happened to boldness in defense of the earth?"[22] Brower is not alone in his criticisms of the Sierra Club. Environmental journalist Frank Clifford argues that "among environmental groups, the Sierra Club may be the best example of bureaucratic heft. With its $40 million annual budget, its Washington lobby and legal arm, its $6 million book business, and $3 million worldwide eco-tourism operation, the Sierra Club looks more like a holding company than a society for nature lovers."[23]

Ironically, thanks to its process orientation, similar messages have emanated from the grassroots. At the peak of the Club's budgetary problems, there arose a dissident movement from within its grassroots ranks, the Association of Sierra Club Members for Environmental Ethics (ASCMEE). Grassroots members in Montana, northern California, and elsewhere in the northwestern United States became disillusioned with the "deal-making between (national) Sierra Club leaders, Congress, and the timber industry."[24] The perceived "inside the beltway" compromises made by the national leadership served to activate many members to redirect the Club's policy agenda; according to George Russel, an ASCMEE leader, "Keep in mind that I love the Sierra Club. All they have to do is have guts."[25] While these messages from the grassroots were registered at the national headquarters, they often left the leadership perplexed. In 1994 internal hostilities reached a point of leaders threatening to cut off computer access to members who were using electronic mail to vilify them. "We used to wake up in the morning knowing who the enemy was and how to face him," said Carl Pope. "All too often lately, we seem to be treating each other like the enemy."[26]

It is even more ironic that the Sierra Club is characterized as part of the "inside the beltway" Washington community as the Sierra Club remains one of the few leading public interest organizations that has its headquarters outside of the nation's capital. McCloskey, however, argues that the organization remains in San Francisco for reasons totally unrelated to policy effectiveness. The issue of relocation was raised at the leadership level in 1985, but he "found no constituency within the organization anywhere to look at the question. The staff wanted to stay. Board members loved to come here for meetings. The Council loved to come here for meetings, too. And, the Washington staff didn't want to have the rest of the staff in their hair back there."[27] More than a decade later, the Sierra Club headquarters still remain in San Francisco, with no move on the horizon. In fact, the location of the national headquarters more closely represents the geopolitical realities of organizational maintenance of the Sierra Club. According to an internal analysis of the geographic distribution of membership in 1991, almost one-third of all Sierra Club members live in California.[28] Little has changed in the regional distribution of members since the study was completed. In the meantime, the Washington presence of one of the largest national environmental organizations amounts to a comparatively cramped office near Capitol Hill.

In many ways the Sierra Club's financial crisis that began in 1991 focused attention on the organizational aspects of the Club that had become "unhitched." By mid-1991, Michael Fischer, who had been hired from outside the Club, announced his intention to leave his position as executive director. He was replaced on October 26, 1992, by a longtime Sierra Club staffer and the former assistant executive director, Carl Pope. While the transition should have occurred in a more timely fashion, the choice of an internal candidate with almost two decades of Club experience was a good one. In the midst of the organizational turmoil that included staff reductions, budget cuts, and a reordering of processes and priorities, Pope provided the stability and institutional memory that served the Club well in this most difficult time.

Many of the problems that plagued the Sierra Club as well as a wide variety of public interest organizations were beyond the control of the organizations' leaders. It is important to note when analyzing the broad economic and political conditions that the financial problems faced by many national environmental organizations predate the 1992 presidential elections. Many environmental groups saw their memberships and budgets peak in 1990 or 1991 and were already suffering from budget shortfalls by November 1992. The nationwide economic downturn of that time had its negative consequences for the entire nonprofit sector as patterns of charitable giving by individual donors tend to mirror the economic conditions in the United States. When economic times are difficult, discretionary spending in the form of charitable contributions tends to fall. In 1991, as a result of the economic recession, charitable giving declined nationwide.[29] Many national environmental organizations felt the resulting economic pinch.

Beyond the broad economic climate, the Sierra Club was also afflicted by an ill-timed membership drive. Its 1991 national direct-mail campaign took place in the midst of the Persian Gulf War; "no one paid attention," said Carl Pope.[30] Another factor that influenced the financial strength of the Sierra Club in the late 1980s and early 1990s and continues today is the increased competition for members, not only among environmental groups but among all other public interest organizations. The growth of animal rights organizations in the late 1980s has had a negative impact on the recruitment efforts of a number of national environmental organizations. In addition, the broad patterns of charitable giving in the 1990s show significant increases in the international affairs sector, particularly in donations to organizations providing direct food and medical relief to starving people in Somalia, Rwanda, Bosnia, and elsewhere around the world.[31] Such changes in donation patterns often result from the shifting of donor dollars from one cause to another, rather than from dramatic increases in the number of donors or in the amount and number of individual donations.[32]

The election of Bill Clinton and Al Gore in 1992 certainly did not serve to activate environmental activists. The success of the Democratic Party in capturing the presidency and maintaining control of both houses of Congress in 1992 lulled many environmentally aware citizens into a state of complacency. The national media was comparatively inattentive to environmental issues as well. Carl Pope in 1994 lamented the significant decline in media coverage of environmental issues following the 1992 elections and the virtual absence of environmental discussion in the 1994 congressional election campaigns (as presented in tables 2.4 and 2.5). "A major problem we face is that media coverage of the environment has declined by some 50 percent in recent years. What has suffered has been the exposure of the very environmental threats we are up against. . . . There is no doubt that if candidates for public office saw more environmental coverage in the media, or if they were challenged on the issues at news conferences and in public forums, we would hear more about the environment in campaigns."[33]

The joint forces of the economic recession, organizational competition for resources, the Clinton-Gore election in 1992, and decreased media attention served to force Sierra Club leaders to "retrench, regroup, and rethink" their entire operation.[34] According to Pope, "we need to get back to basics." "We don't have the dollars that we need, even at this time of tremendous need. We will simply have to use the dollars that we have even more effectively, something which our network of volunteer activists and experienced staff enables us to do with tremendous impact."[35] To that end, J. Robert Cox, then president of the Sierra Club, announced in late 1994 "Project Renewal," the Club's program for organizational reform: "Difficult as these decisions are, we had extraordinary Board consensus that we had to consolidate our staff size and structure if we were to reduce expenditures and improve our organizational efficiency. We know that our members want us to plan responsibly and take action to fulfill our mission."[36]

In the midst of the Club's restructuring and redefinition, the Republican successes in the 1994 congressional election served as a catalyst to reenergize the Sierra Club and the environmental movement more generally. Just as Ronald Reagan and his administration provided tangible targets for opposition, so, too, did the Republican leaders in Congress, particularly in the House of Representatives. The resulting recruitment and mobilization strategies targeted at these newly identified "enemies," undertaken by the Sierra Club and other organizations, have proved successful in 1995, 1996, and 1997 as memberships and budgets are beginning to rebound from the declines and deficits of the early 1990s.

The net result of the organizational restructuring is a more coherent and focused Sierra Club, less constrained by the consensus-driven, process orientation that inhibited the abilities of professional staff and leadership to act on behalf of the membership in an expeditious manner. While the Sierra Club continues to maintain a complex structure that links members with national leaders, its new streamlined decision-making process, brought about by a pragmatic, bottom-line assessment of its mission and operation, will serve the Club well in years to come. The economic realities of the early 1990s forced the Sierra Club leadership to reevaluate not only the structure of the organization but its mission as well. The reconstituted Sierra Club in many ways is better equipped to do more with fewer resources. The Club's new "ACT" initiative has the potential to provide a more meaningful linkage to the grassroots membership than any of the past efforts at providing a democratically representative decision-making process. ACT, as introduced by former Club President Cox, has three main components: "A" stands for "activist culture," or the effort to involve greater numbers of members in the political process; "C" is for "communication and coordination," a push to help Club volunteers take advantage of new information technology; and "T" is for "training," primarily in grassroots organizing skills. "To the degree that we invest in healthy groups and chapters and training for grassroots effectiveness," says Cox, "the Sierra Club's national priority goals also will be met."[37]

Of the five organizations discussed here, the Sierra Club has the organizational structure most conducive to linking its organizational *voices* with its *echoes*. The Club's new "re-hitched" structure redirects the organizational focus away from the internal workings and toward the policy-making process. It provides greater opportunities for more meaningful activism between leaders and members and offers leaders greater flexibility to refine and adjust their voices in advocating environmental positions. The election in 1996 of a Generation-X president, Adam Werbach, served to rejuvenate and mobilize the Sierra Club membership and to attract a younger base of support for the years to come, but not without some dissent. Werbach did not escape the wrath of the press, having been characterized as simplistic, superficial, and even silly. His public statements caused unease among longtime members of Sierra Club. For example, Werbach stated that he does not want the Sierra Club thought of as an "organization of aging hippies." Demographically, the Sierra Club membership, in fact, includes many who were politically socialized in the 1960s. Nonetheless, his symbolic leadership had a demonstrable effect on refocusing the Sierra Club on its grassroots and on a new, younger audience of potential environmentalists.[38] The subsequent election of Charles McGrady as president in 1998 has returned the Sierra Club leadership ranks to the more traditional demographics of current national environmental leadership.

The Wilderness Society: The Leadership Roller Coaster

To understand fully the leadership trials and tribulations that have defined the character of The Wilderness Society, one needs to venture back more than thirty years into the organizational archives of the Society. In 1965 The Wilderness Society celebrated its thirtieth anniversary with its 20,000 members. In the years that followed, with the increased attentiveness toward environmental issues and the beginning of the third wave of environmental activism, existing conservation organizations such as The Wilderness Society, the Sierra Club, and the National Wildlife Federation experienced dramatic increases in their memberships. By 1970 The Wilderness Society had more than 50,000 members (see table 2.7).

As The Wilderness Society grew through the second half of the 1960s, its activism and effectiveness were largely attributable to the efforts of the executive secretary, as the director of the Society was called at the time. Two men, Howard Zahniser and Stewart Brandborg, led the organization through the 1960s and into the 1970s. However, with the growth of membership, budget, and staff in the 1970s came organizational disarray; there was no leadership control over the growing enterprise. As a result, Brandborg was fired in 1976. For the next two years the Society continued to flounder under two additional executive directors. By 1978 the organization was on the verge of bankruptcy. Payrolls were not being met; membership development had lapsed, and the communications to members were curtailed—then came William A. Turnage.[39]

William Turnage assumed the position of president of The Wilderness Society in November 1978. At the time, there were thirty-seven employees in the organization. By Christmas, thirty-six of these employees were seeking other jobs. Turnage fired all but one of the existing staff. His initial assessment of the organization was that it was being run just like all the other public interest organizations at that time. "There has been an unfortunate tendency in environmental organizations because the staffs tend to be made up of people who, if they are not alternative lifestylers, they have opted out of the corporate, structured, managed sort of rat race. They don't want people telling them what to do. They want to do their own thing."[40] As he was climbing aboard the sinking ship, he was not at all interested in appeasing the interests of individual staffers; he wanted to lead the organization and to manage it in a professional, businesslike manner. His first order of business was to restructure the entire organization.

> They [other groups] are not hierarchical organizations. They are not structured in the conventional business or governmental sense. They are congeries of individuals who are working on projects that are of interest to them, generally funded by the organization, and that's the main function the organization plays. Here at The Wilderness Society, we are structured in a hierarchical way, and we are run very much like a business or governmental agency, hopefully less bureaucratically and hopefully with a little more sort of modern compassionate kind of management where people count a bit more. But, the priorities and strategies of the organization are ultimately decided by one person; whereas, in the other organizations, if they have them at all, they are consensus-oriented. *And consensus invariably achieves a more mediocre product than leadership achieves.* These organizations are trying to change society by leading the way, by having bold, imaginative ideas and by being spearheads. Well, if your internal management is even more conservative than big business or government because it's all consensus, which is inherently ultraconservative, how are you going to lead? There is certainly no question in my mind that The Wilderness Society is the best managed of any of these groups, [and] I'm the best chief executive in any of these organizations. It is partly because I am willing to lead.[41]

Turnage served as president of The Wilderness Society until December 31, 1985.[42] During his tenure, Turnage more than doubled the organization's budget and more than tripled its membership. As is evident from his statement cited above, Turnage was quite frank in his comments regarding his organization as well as the other organizations in the national environmental movement, but he was also philosophical, taking stock of what he had accomplished and what was left to be done. Turnage outlined three factors necessary for organizational viability in the public interest sector: structure, leadership, and management. Regarding the organizational structure, he argued that while it is not necessary to have an organizational infrastructure that reaches down to the grassroots, it is necessary to link leaders and members in some significant way.

> There has to be a connection between the two. You cannot survive for long, that is you cannot maintain for long, if you are not serving a constituency. Now, how you relate to that constituency is something totally different. You have the Sierra Club that is, to an extraordinary degree, connected to its grass roots and to many of its members, but they are the most powerful force in the Sierra Club. A group like ours, we don't have the chapters, we don't have *that* structure, but I would argue that much of the success we have had in the last few years is because we are connecting to a constituency—we are serving a group, even if it is in a totally different way than the Sierra Club does as far as the structural configuration they have. The only way you can be successful in a conservation organization, and it's probably true of public interest groups in general, is if you are very definitely serving a constituency. You can draw an analogy to Mercedes-Benz, the most successful, most profitable automobile manufacturer in the world. They are clearly connecting to a market.[43]

According to Turnage, the existence of a complex infrastructure that incorporates grassroots input is neither a necessary nor a sufficient condition for political representation in the public interest sector. Internal relationships between board and staff are also evaluated in a nontraditional manner. Turnage contends that board members serve as "guardians of the long-term philosophical integrity of the organization." As such, board members of The Wildnerness Society play more indirect roles in the policy-making process. Robert Blake, a member of the Society's board of directors, summarizes the board-staff relationship: "The board will help set the policy, but it will not interfere with staff. The board members will never tell staff members what to do. . . . The board is basically a check, a watchdog for the long-range interest of the organization. Initiative has to lie with the staff; that's what you have a staff for."[44] As a result, the Society was organized with one goal in mind—effective policy advocacy in a clearly defined policy area.

Policy effectiveness, however, requires more than a staff-dominated organization; it requires leadership. The leadership provided by Turnage during his seven-year reign may be classified as "transforming." James MacGregor Burns states that transforming leadership "occurs when one or more persons *engage* with others in such a way that leaders and followers raise one another to higher levels of motivation and morality."[45] Turnage offered his philosophy on public interest leadership:

> This is a political business. . . . I believe in the notion of leadership. Nobody dislikes Ronald Reagan more than me. But, the god-damned guy is a leader. He really makes decisions. He's tough; he's charismatic—he understands power. We never had a president so ignorant of the issues, so ignorant of substance, but, boy, does he understand power—how to use it and how to apply it—when to push, when to pull back . . . fantastic. Now if you're in a political business, if your business is creating new perceptions, if your role is to be a change-maker, an educator, a stimulator, you've got to have leaders—people who personify this, who can present an issue in a dramatic way, who get people's attention, who can focus an organization's energies so that it doesn't just piss around, it actually gets something done.[46]

Organizational leadership, in addition to staff management skills, requires strategic planning on the part of the organization's leaders. There is no strategy involved in having thirty staffers, each doing his or her own projects. A unified effort is necessary for maximum effectiveness. Turnage argued that strategy is the critical element to effective leadership. "A lot of people confuse strategy and tactics. Tactics is what happens in the battle; whereas strategy is how you win the war. And, you've got to have a strategy for your institution. Frankly, I doubt whether many of these people do. There isn't necessarily any one strategy that's right, but, boy, you better have one and you better have only one."[47]

The Wilderness Society, under the direction of Turnage, employed a four-point strategy: (1) make policy focus discrete and clear, a very defined set of issues—public lands; (2) use the media—advertising and credit-claiming, making the agenda public; (3) use the techniques of business management and apply them to policy advocacy—professionalization; and (4) become known as the most critical and outspoken organization in the industry. According to Turnage, The Wilderness Society had implemented a successful political strategy during his tenure.

The organization did not become an "environmental conglomerate," dealing with issues as diverse as clean air, toxics, population, nuclear and solar energy, and wilderness issues, as did some organizations in the industry; rather, it was focused on a limited set of issues. At the time of Turnage's departure, the Society had more professional media staff than the top ten environmental organizations combined. Regarding management practices, Turnage made all employees dress in a professional manner. He argued that "you can kick the other guys twice as hard in the shins if you are wearing a coat and tie and know what you are talking about, because you have got credibility."[48] Finally, the Society sharpened its image as a colorful and hard-hitting public interest organization.

Another important factor that influences organizational effectiveness, according to Turnage, is management. The Wilderness Society was the first public interest organization to overhaul its entire governing infrastructure. After the thirty-six employees were fired in 1978, new criteria for employment were imposed. No longer did it matter whether one marched on Earth Day or had protested in Chicago or marched on Washington. Nor did it matter that one was good friends with the conservation director. New staffers were selected based on professional skills and policy expertise. Today, the more than one hundred professionals working in Washington or in the regional offices are representative of the new breed of environmental policy advocates—well-trained, well-paid policy professionals. More than any other aspect of the organization, Turnage was most proud of the Society's success as a professionalized policy advocate.

> I think the professionalism of the Society and the tremendous knowledge of the staff are perhaps our strongest assets and we try to use these capabilities in a *non*-confrontational way—but as *advocates*. . . . The conservation movement has changed and has become very sophisticated, and I think we're the most sophisticated organization in the movement. We're trying to deal with government and business not as minor supplicants but, instead, on an equal footing, using similar levels of expertise and knowledge.[49]

William Turnage served in a variety of crucial capacities in the resurrection of The Wilderness Society; he was a political entrepreneur, a leader, a manager, a salesman, an intellect, and a man of courage and boldness. In fact, through his efforts, political entrepreneurialism has become less important as a criterion for leadership in the Society. His legacy to The Wilderness Society is the creation of a professionalized organizational infrastructure that needs to be managed and directed, not coaxed and cajoled. The choice of his successor addresses this legacy.

Turnage's successor at The Wilderness Society was George Frampton. Hired from outside the organization, Frampton came to The Wilderness Society from the legal profession. He was one of the Watergate prosecutors; he had also worked on the Three Mile Island investigations. In the Society's official announcement of Frampton as the successor to Turnage, Frances Beinecke, former chairperson of the Society, proclaimed, "We have found a first-class leader. George Frampton is blessed with advocacy skills, a prosecutor's toughness, outstanding intellect, and integrity. He is committed to the public interest and to the preservation of our public lands. Mr. Frampton is a man who commands respect, and we are very pleased that he has agreed to take on this challenge."[50]

With his solid credentials, Frampton was an able successor to Turnage, but he was quite different in style and organizational outlook. At the time of the internal decision making that led to the selection of Frampton, Sharon Dreyfuss, membership associate for the Society, noted that Frampton "is not Bill Turnage, but no one can be Bill Turnage but Bill Turnage."[51] Frampton more clearly served as a transactional leader, managing the organization through a time of growth in the environmental movement. From 1986 to 1990, The Wilderness Society flourished; its membership jumped from under 150,000 members to 350,000 members and its budget almost tripled, rising from $6.3 million to $17.9 million.

Like the Sierra Club and a host of other national environmental organizations, The Wilderness Society also felt the economic pinch of the early 1990s. Through 1991 and 1992, the Society initiated a series of cost-cutting efforts that resulted in a decrease in expenditures of more than $1.5 million, which helped reverse the $1 million deficit in 1991 even as revenues dropped by more $200,000 in 1992. During this time its membership fell by more than 50,000. In the midst of the leadership responses to the growing maintenance problem, Frampton stepped down as president of the Society, as he was nominated to be assistant secretary of the interior on February 23, 1993.

Vice President for Conservation Karin Sheldon served as acting president of the Society, replacing Frampton until G. Jon Roush was appointed president in October 1993. In Roush, The Wilderness Society found another transformational leader, not unlike Bill Turnage. Roush was equally, if not more, controversial. He also had a vision of public interest leadership and a strategy to implement his vision.

Roush, formerly president of Canyon Consulting, an environmental management and planning firm, came to The Wilderness Society with significant environmental credentials. During the 1970s and 1980s he had served in a variety of leadership capacities at the Nature Conservancy, including executive vice president and chairman of the board. Prior to his career in environmental advocacy, he had taught at the college level. He has written for a variety of environmental publications and is the author of *The Disintegrating Web*, published in 1991. In addition to these pursuits, Roush also owns a ranch of almost 800 acres in western Montana, valued at more than $2 million.

Prior to assuming the leadership responsibilities of The Wilderness Society, Roush presented his vision of environmental leadership in the 1990s.[52] He argued that contemporary environmental leaders face two main obstacles in framing a cohesive vision of environmental advocacy: "the differences in style, structure, goals, and working environment among conservation groups and the insularity of conservation organizations within society as a whole. Overcoming these obstacles is a function of leadership."[53]

Roush offered six steps for environmental leadership to achieve a cohesive environmental vision:

1. *Become obsessed with communication.* Look for or create opportunities to communicate laterally or vertically with other leaders.

2. *Create, support, and take advantage of opportunities for leadership training.*

3. *Think strategically.* Start by asking the following questions: (1) What can your organization do that is useful for people as well as the natural world? (2) What can you do that is unique? (3) What can you do to establish the highest standards of quality? (4) How can you become recognized for doing something useful, unique, and excellent?

4. *Act for the movement.* Share what you have learned with leaders of other organizations. Turn competitors into collaborators. Believe that you gain power by sharing it.

5. *Broaden the movement by broadening your organization.* If your supporters include few people from ethnic minorities, few women, or few poor people, find out why. By all means, look for ways to bring them in.

6. *Broaden your organization by striking new kinds of alliances.* Everyone has a role to play. Business contributes technical expertise and decisions about the use of natural resources. Government contributes societal goals and coordination. Nonprofits contribute a concern for externalities, education, and vision.[54]

Roush concludes his essay on leadership in much the same way that Turnage would have: "For a conservation leader in the late twentieth century, the organization is the only tool available. The leader carries the vision. The organization supplies the voice."[55]

The leadership model offered by Roush bears the imprint of his experiences at the Nature Conservancy, particularly as it relates to organizational maintenance. Today, the Nature Conservancy is, by far, the wealthiest environmental organization in the United States (see table 2.6). Fund-raising efforts at the Nature Conservancy are diversified to include not only direct-mail solicitations for individual member donations but also appeals for institutional support from foundations, corporations, and the government through grants as well as the courting of high-dollar donors.[56] Roush attempted to diversify the fund-raising efforts at The Wilderness Society in order to compensate for the comparatively low annual membership rate of $15.00. His efforts, accompanied by the closing of five field offices and a 20 percent reduction in staff, raised questions regarding the vision of Roush becoming too focused on advocacy "inside the beltway" at the expense of activities throughout the United States. According to journalist Tom Kenworthy: "Despite its pedigree in the environmental community, The Wilderness Society in recent years has been criticized by more militant environmentalists, who say it is too willing to compromise and too comfortable in Washington's corridors of power."[57]

With former president George Frampton at the Department of Interior, board member Jim Baca directing the Bureau of Land Management at the beginning of the Clinton administration,[58] and former board chairperson Alice Rivlin serving as vice chairperson of the Federal Reserve Board, one might expect the Society's focus to be on Washington. Nonetheless, Roush himself became the target of militant, grassroots environmentalists' attack on "corporate environmentalism," when it was made public in April 1995 that Roush had sold 400,000 board feet of timber (roughly 80 acres, valued at $140,000) from his Montana ranch. Environmental journalist Alexander Cockburn, who had attacked The Wilderness Society for the corporate representation on its governing board in his "Beat the Devil" column in the February and March 1995 issues of *The Nation,* joined with Jeffrey St. Clair in the April 24 issue of the magazine to criticize Roush for betraying the ideals of the organization: "The head of The Wilderness Society logging old growth in the Bitterroot Valley is roughly akin to the head of Human Rights Watch torturing a domestic servant."[59]

Roush responded by presenting the timber sale as a "model of good forestry practices that was supervised by a forest consultant and far exceeded the requirements of Montana's forest practices law and that no old-growth trees were cut. 'Nothing over 90 years was taken.' Roush said he decided to sell the timber only after he was unable to sell his ranch to pay for a divorce settlement and tax bills."[60]

In March 1996, having been unsuccessful in weathering the timber sale storm among other internal controversies, Roush left The Wilderness Society. As the search for a new president took place, Mary Hanley, formerly vice president for public affairs, worked with the governing council to manage the day-to-day operations of the organization under less than optimal financial circumstances.[61] After months of searching for a new leader, William H. Meadows III was selected by the governing council as the Society's new president in late September 1996. Like Roush before him, Meadows came to The Wilderness Society with established fund-raising credentials in the national environmental community, having spent the previous four years directing the Sierra Club's capital fund-raising efforts. Meadows began his presidency in early December 1996.

Despite the recent membership increases and the expectation of new executive leadership, the expenditures for *Wilderness* magazine forced the editors to reduce the quarterly offering to twice a year, beginning in April 1996. Roush failed to transform the organization, due in no small part to his public relations debacle. The Wilderness Society's governing council hopes that Meadows will serve as a leader who possesses the skills to clarify the mission, broaden the audience for support and activism, and provide the vision necessary to strengthen the organization for the next century.[62] During his first six months in office, Meadows has maintained a low-key approach, rarely commenting publicly on environmental issues on behalf of The Wilderness Society.[63]

National Wildlife Federation: Environmental Conglomerate

Leadership at the National Wildlife Federation entails the management of a roughly $100 million educational, merchandising, and publishing enterprise, the maintenance of a membership base of more than half a million associate members and more than four million citizens affiliated with state and local chapters, as well as the representation of the organization and its membership in the policy-making process. At first glance, the federation has a comparatively representative organizational structure. Upon closer examination, however, one finds no more than the facade of a representative process. According to General Counsel Joel Thomas, "the Federation was put together like a labor union, and it was no accident that was the model. Individuals joined clubs; clubs joined either state organizations or subdivisions of state organizations. And the state organizations affiliated with the national organization."[64] Thomas further states that the control of the national organization is ultimately in the locals, although it is difficult to demonstrate that in this particular organization.

The National Wildlife Federation claims to have more than four million members across the country connected at some level with a Federation-affiliated organization. However, according to John Gottschalk, member of the board of directors of the National Wildlife Federation:

> Our relationship with the bulk of the membership is very tenuous. It probably doesn't exist except in an almost imaginary way. We know from time-to-time pretty much what's going on because we do professionally managed surveys to get their opinions on a variety of issues, and that has been extremely useful. But, by and large, I think it would be fair to say that the four million members of the world's largest conservation organization is just a great mass of subscribers to magazines. And out of that number, just a handful are activists. A handful in terms of those numbers, now that may be 25,000 people who are willing to sit down and write a congressman when asked to do so, but that's enough to have a great effect on Congress.[65]

Federation leaders are not troubled by this relationship because the policy-making process established by the leadership does not include direct grassroots input. Joel Thomas explains the Federation "has another type of policy-making system that does not require us, in effect, to keep our members as apprised or to encourage their interest in specific policy areas, because the policy is going to be set through the affiliate organizations."[66] Despite the tenuous relationship between members at the grassroots level and the national leadership, there are, indeed, structural linkages that connect state and regional organizations with the national Federation. Approximately half of the twenty-six members of the Federation's board of directors are elected from the various regional components of the organization; the remaining positions are filled through appointments by the board. The board hierarchy includes a president, historically someone who has risen through the affiliate ranks, and three vice presidents representing the eastern, central, and western regions of the Federation.

While the governing structure of the Federation is not nearly as complex as that of the Sierra Club, the board does provide some direction for the professional staff. John Gottschalk summarizes the role of the Federation's board and provides an interesting anecdote relating to board operations.

The board functions in several ways: actually it does have the opportunity to initiate new programs or changes in existing ones, but, for the most
part, the board acts as a sounding board for the executive staff and reviews programs in the regular budget setting process which takes place at
the summer board meeting each year. . . . A few years ago, there was a
feeling that the Federation was becoming overly activist, and, of course,
we all see the Federation as a middle-of-the-road group. One of the
most difficult tasks that the executive staff has is balancing the noncon-
sumptive conservation constituency across the country, the protectionist
element, shall we say, against a fairly vocal and strong consumption-oriented element, the hunters and anglers, and so this is very much on their
minds. There was a feeling on the part of the consumption-oriented
people that the Federation was getting too far afield in environmental
protection and litigating too much and had too damn many lawyers on
board. And this got to be a big issue at one of our summer meetings. It
was rugged. There were occasions when we asked people to leave the
room so we could have an executive discussion and call the S.O.B.s what
they really were, face-to-face. It was probably the most contentious
board meeting that I ever attended. But we did it. We thrashed it all out.
The charges were leveled. The defendants brought before the bar, and
found innocent as a matter of fact. The board voted to support the direction that the Federation was going, with the result that that particular element of the board either did not get reappointed or reelected by the
board, or dropped out. That's the way things like that work.[67]

The National Wildlife Federation is unique among contemporary envi-
ronmental/conservation organizations in its approach to its membership.
In many cases, Federation members are little more than subscribers to
magazines. The political connection between these individuals and the national leadership is virtually nonexistent. The communications received by
these subscribers have comparatively little specific policy content, as will be
demonstrated in chapter 6; therefore, the organizational structure as well as
the communication process serve purposes other than providing some representational link between leaders and members.

In fact, the magazine publishing and merchandising enterprises of the
National Wildlife Federation generate substantial profits. As reported in
chapter 2, fully two-thirds of the Federation's revenues are generated
through sales of merchandise and magazines. In 1994 these revenues produced $26 million in profits for the organization, resulting from markups
of up to 200 percent on some products.[68] In comparison with the National

Wildlife Federation, the Sierra Club generated only $1.5 million in profits from its merchandising and publishing enterprises in 1994, while Greenpeace produced profits of $2.3 million from its merchandising.[69] Given that the Federation and its foundation, NWF Endowment, Inc., are 501(c)(3) nonprofit, tax-exempt organizations, many journalists and other interested parties have questioned the legality of such profit-making enterprises within so-called nonprofit organizations.[70]

The policy-making apparatus of the Federation does include input from state federations through the representation of these entities on the governing board. However, as described by Gottschalk, the organization is very much a staff-dominated operation, with the board assuming a watchdog role in terms of policy outputs. Even at the annual meetings, where state federations may present policy resolutions for adoption by the entire organization, Federation staff provide commentaries and recommendations in order to inform voting participants of the position of the professional component of the organization. Staff recommendations are rarely overturned. Despite this staff dominance, board and staff consciously seek a consensus on the direction of the policy agenda. Compared with the Sierra Club, however, the policy-making process is more concentrated and less reliant on consensus-building mechanisms. Grassroots participation in the policy-making process does not exist. In most cases, members (subscribers) cannot possibly know the policy agenda of the Federation.

While the organization has a broad policy agenda that encompasses the traditional conservation issues as well as the more technical environmental concerns, its development is largely undertaken at the national level. According to former President Jay Hair, "We must find new and creative ways to develop policies that guide our public and private institutions. We must integrate such varied disciplines as science, engineering, economics, and law as we seek solutions to a wide range of complex issues before society."[71] During his fourteen years as president and CEO of the National Wildlife Federation, Jay Hair sought to manage the organization as an environmental conglomerate, wherein the three goals of managing the merchandising and publishing enterprise, maintaining the membership base, and representing the political interests of the organization were, from a management perspective, separate and distinct. As a result, policy formation in the Federation did not necessitate the direct input of members. By the early 1990s, Hair had built the policy division of the Federation into one of the most comprehensive research and advocacy efforts in the environmental movement. However, due to the size of the overall budget, the federal lobbying activities account for less than 1 percent of the organization's annual expenditures.

In an effort to balance the policy perspectives of the Federation, Hair, a fervent believer in bringing the environmental community together with business leaders in order to solve environmental problems, reached out to the corporate community to initiate a dialogue. He also reached out to business interests for financial support by creating the National Wildlife Federation Corporate Conservation Council in the early 1980s. For $10,000, major corporations could join the council; by 1990 corporations such as Monsanto, DuPont, Waste Management, Inc., and Ciba Geigy made contributions to the Federation Council that totaled more than $400,000.[72]

The National Wildlife Federation is not alone in its pursuit of corporate financial support. The World Wildlife Fund has accepted major donations from Chevron, Exxon, Philip Morris, Mobil, and Morgan Guaranty Trust, while annual corporate contributions to the National Audubon Society surpassed one million dollars by 1990.[73] Still other environmental organizations, such as the Sierra Club, accept corporate advertising revenues through their magazines. Beyond the acceptance of corporate contributions, Hair and the Federation added several corporate executives to the Federation's board. In 1987, for example, Dean Buntrock, head of Waste Management, Inc., one of the largest operators of incinerators, landfills, and hazardous waste sites in the United States, joined the board at Hair's invitation.

Unfortunately for Hair, his willingness to reach out to corporate America was viewed by some in the environmental movement as being far too accommodating. According to environmental journalist Art Kleiner, "He's built his career around getting people to trust him. Yet, to judge from his press clippings, he could be the least trusted environmentalist in America."[74] In addition, his management and personal style within the Federation began to raise questions regarding the changes in the organizational culture. The new Washington headquarters, completed under Hair's direction in 1990, had the look of corporate offices, complete with mahogany desks. His lifestyle also tended toward "business class." During his final year at the Federation, Hair received a salary and benefits package of more than $325,000. Furthermore, his travel budget swelled from his frequent trips to Europe, Asia, and elsewhere around the world, not only on Federation business, but also on travel related to his presidency of the World Conservation Union, an international environmental association.

After months of internal staff conflicts relating to Hair's management style and to his comparatively lavish personal lifestyle, particularly in the midst of declining membership and merchandise sales, the staff staged an "open revolt" in the spring of 1995. According to former Senior Vice President Larry Schweiger, "He had been absent without leave."[75] After fourteen years as president of the National Wildlife Federation, Jay Hair resigned voluntarily on July 3, 1995, "to devote more time to his family and to international environmental affairs."[76] Hair remains affiliated with the Federation as "President Emeritus" and continues as president of the World Conservation Union. Former Executive Vice President Bill Howard replaced Hair as president and CEO of the National Wildlife Federation. Like Meadows of The Wilderness Society, Howard maintained a low public profile during his brief tenure between July 1995 and June 1996, rarely speaking on behalf of the Federation.

The National Wildlife Federation was not the only national environmental organization to experience a major management reorganization in July 1995. The National Audubon Society, also suffering from membership losses and declining revenues in the early 1990s, hired a management consulting firm to assess the operation of the organization. As a result of the assessment, the Audubon board replaced longtime President Peter Berle with John Flicker, formerly of the Nature Conservancy, on July 5, 1995. Today, both organizations are in the process of reorganizing themselves and redefining their missions for the next century.

The National Wildlife Federation, in particular, has sought a significant "image makeover." After the brief tenure of Howard, the Federation board unanimously voted to elevate Mark Van Putten, a forty-three-year-old veteran grassroots organizer and director of the Federation's Great Lakes Natural Resource Center, to the position of president. Rather than the more than $325,000 salary and benefit package enjoyed by Hair, Van Putten is paid $170,000, only slightly more than the average salary for heads of national environmental organizations identified in the analysis presented earlier in the chapter.[77] One of Van Putten's first acts as president was to sell off the Federation's interest in the palatial downtown Washington headquarters, which had been recently renovated by Hair, and move the staff to its suburban northern Virginia facility, already owned by the Federation.[78]

Most recently, Van Putten has sought to sell off the forty-three-acre tract of land that includes the new headquarters property. The land with the buildings, assessed at $12.8 million, is prime real estate in the crowded Washington suburbs. He then hopes to relocate the Federation headquarters to a smaller, more economical space.[79]

Van Putten recognizes that in addition to shedding the corporate image through recasting the physical appearance of the Federation headquarters, he must also reinvigorate the grassroots infrastructure of the organization. He has identified the break between the Federation's Washington operations and its grassroots support system across the country and is seeking to remedy the problem. Van Putten brings to the Federation a brand of pragmatic leadership that should serve the organization well in the years to come.

Environmental Defense Fund: Market-Based Environmentalism

Representative of the new era of environmental advocacy in the United States, the Environmental Defense Fund was established in response to a particular environmental disaster, the poisoning of ospreys with DDT on Long Island, New York. The initial success of the small group of scientists and lawyers led to the creation of an organization that "links science, economics, engineering, and law to create innovative, economically viable solutions to today's environmental problems," the Environmental Defense Fund (EDF). The organization literally began with an idea and a new approach to environmental advocacy. What it needed was the necessary funding to implement the idea and approach. Through foundation support and donations from philanthropic organizations, EDF was able to establish an identity in the new environmental arena.

Unlike its older conservation counterparts, EDF had no real social network through which to build constituency support. It had to rely on the memberships of related organizations to build a contributor base of its own through direct mail—a strategy to be analyzed in detail in the next chapter. As a consequence, the relationship between leaders and members is qualitatively different from that in the movement's older organizations. There is no organizational structure that links the grassroots to the leadership. While EDF, headquartered in New York City, does have regional offices across the country in Oakland, California; Boulder, Colorado; Raleigh, North Carolina; Austin, Texas; and Washington, D.C.; as well as a project office in Boston, these offices do not serve as regional coordinating centers for members. In fact, EDF leaders look upon their support base as contributors rather than members. As a result, they do not attempt to mobilize their contributors as volunteers to sign petitions or write letters to members of Congress. According to Frederic Krupp, executive director of EDF, his supporters are making their statements about the environment through their contributions: "They have their own lives and their own concerns, and, as such, the large majority of people who do belong want to be environmentally responsible. And, one of the major ways they are environmentally responsible is by donating as much as they can to the Environmental Defense Fund, and perhaps a range of other environmental groups."[80]

Organizational governance, as a result, is left to Krupp, the professional staffs, and the board of directors. Unfortunately, the decentralized organizational design of EDF makes effective governance a difficult task. With its regional offices throughout the country, each conducting research on different (and similar) environmental issues, along with the research wing of the New York headquarters also conducting policy research, EDF has remained one of the more loosely organized national environmental organizations. Even the administrative apparatus is dispersed. General administration and accounting is conducted in New York, yet the membership and development office is in Washington, D.C. Similarly, the director of communications is in the capital, while the *EDF Letter*, the organization's primary method of communication with its membership, is published in New York.

Krupp admits that decision making is decentralized but supports such a process on the following grounds: "All the decisions are not made here in New York; all of the people in the regional offices are incredibly brilliant and able. There is the presumption that when they recommend some action, they would not do that unless there was good reason."[81] The board of trustees acts to promote this method of operation. David Challinor, EDF board member, argues that "[only] a small percentage of the board is equipped to make policy judgments, the Executive Committee. Nonetheless, there is a great deal of discretion given the permanent staff; therefore, the relationship between the staff and the board is important—mutual trust and confidence."[82]

Regarding its organizational decision-making structures and processes, my initial assessment of EDF was less than positive: "The Environmental Defense Fund is a textbook example of organizational ambiguity. On all dimensions, EDF lacks the organizational structures and processes for effective governance."[83] While little has changed structurally in the last decade, EDF has matured as an organization and has adapted its rather fragmented organizational style to the demands of the late 1990s.

One of the reasons for the success of EDF in the early 1990s, a time when many environmental organizations were floundering, is that the organization was late in adopting the recruitment and marketing strategies used by other national organizations. While its primary method of membership solicitation has been direct mail, the membership development wing of the organization was not allocated sufficient resources to be competitive in the direct mail marketplace. As recently as 1986 EDF conducted its recruitment efforts with few of the premiums offered by its competitors. Lacking the proper resources, the organization attempted to market itself as an anomaly. In a 1986 membership renewal letter, Executive Director Fred Krupp offered "No canvas bag, no bumper stickers for continuing with us. Just the knowledge that you will help reduce our country's acid rain by fifty percent." Despite the innovative approach, EDF membership remained at 50,000 through the mid-1980s. Brian Day, hired as the first director of communications in 1985, succinctly identified EDF's problem: "We have been doing good work for a long time, but one thing we forgot to do is tell anybody."[84] Much of EDF's recent success in mobilizing and maintaining 300,000 members after years of membership stagnation at the 50,000 member range may be attributed to creating and facilitating the growth of a communications department in the organization. Between 1985 and 1990 EDF became much more visible as an environmental actor. Its long-running public service advertisements featuring the tag line, "If you're not recycling, you're throwing it all away," were instrumental in raising the organization's name recognition.

Other recruitment efforts generated significant increases in new members. In 1992, for example, EDF gambled by financing an expensive direct-mail recruiting campaign that included a free, full-color EDF calendar; the gamble paid off with thousands of new members. A decade earlier the organization would not have contemplated such a strategy. Finally recognizing that the successful recruitment and, more important, retention of members requires significant expenditures of organizational resources, EDF has shifted its budgetary priorities toward such efforts and has reaped the benefits through the resulting growth in its operating budget and its membership.

Apart from the recognition that membership recruitment and retention required more attention and resources, EDF has also benefited from its unique method and message of environmental activism. During the last decade, EDF has demonstrated its willingness to work with industry representatives in order to identify environmentally sound yet economically feasible solutions to environmental problems. EDF's detractors have labeled this approach free-market environmentalism. "Some environmentalists advocate market mechanisms in the name of efficiency, reasoning that making environmental responsibility cheaper will result in a corollary reduction of political opposition, the end result being the possibility of greater protection. This, crudely put, is the position of the Environmental

Defense Fund, the most market-oriented of the major environmental groups."[85]

Krupp rejects the free-market criticisms: "We are for government-imposed markets, the government use of markets that put economic incentives on the side of innovation, and innovation on the side of defending the environment."[86] Nevertheless, EDF's approach to environmental problem solving has served to elevate the organization to a position of prominence in the contemporary national environmental movement, but not without controversy. For example, in the negotiations that led to the passage of the Clean Air Act amendments in 1990, EDF staffers, especially economist David J. Dudek, worked closely with the Bush White House on several aspects of the proposed amendments, most importantly, provisions regarding the pollution credit trading program. At the same time EDF was a member of the National Clean Air Coalition. As a result of the close working relationship with the White House, other coalition members became deeply suspicious of EDF's efforts.[87] Two years later, as a result of the interaction between EDF and the White House, Frederic Krupp was appointed a member of the Bush Commission on Environmental Policy. At an April 1992 Rose Garden ceremony in which President Bush announced his regulatory moratorium, commission member and guest Krupp decided to take the opportunity to denounce the Bush program. Capturing the microphone Krupp said, "I don't think the [regulatory] freeze will help the economy. And it certainly won't help the environment." He later called the regulatory freeze "a wholesale handout to the American business community."[88] The approach taken by Krupp and EDF has, at times, raised questions from within the environmental community as well as from government officials regarding the strength of the alliances entered into by EDF.

The Environmental Defense Fund has also created controversy within the movement by reaching out to big business in the United States. After doing battle with McDonald's Corporation over the use of polystyrene packaging containers in the 1980s, EDF and McDonald's joined forces in 1990 to create a task force to establish ways to reduce, reuse, or recycle materials generated by McDonald's. At the time McDonald's pledged to spend $100 million annually to purchase recycled materials for construction, remodeling, and operations of its restaurants in the United States. By January 1995, McDonald's had surpassed its goal by spending $1 billion on recycled materials during the first four years of the program. In 1995 McDonald's and EDF joined with the Advertising Council to produce public service advertising to encourage consumers and businesses to buy recycled products.[89] Similarly, in 1992 EDF created a controversial alliance with General Motors to study ways to limit pollution.[90]

These programs, along with the newly created Alliance for Environmental Innovation that will provide further linkages between business leaders and the organization, serve to define the mission of EDF for the next century.[91] The organization's bottom-line approach to environmental problem solving has certainly attracted the attention of the environmentally attentive public, both young and old. According to Krupp, "People want results; they want to get things done, and that's the reason people want to be part of our efforts."[92] Like the Nature Conservancy, EDF is able to present to its members demonstrable results that are attained, in many cases, through means that do not include the use of the federal government. This common method of operation may help explain the success of these two organizations in the 1990s.

Environmental Action: Collective Subsistence-Collective Demise

Environmental Action was organized for the first Earth Day in 1970. Denis Hayes, founder of Earth Day and the first director of Environmental Action, spoke to the Earth Day crowd in Washington more than twenty-five years ago. His comments reflect a view of the environmental movement that was equally applicable to Environmental Action, not only a quarter century ago but even as it closed its doors in late 1996: "We are building a movement, a movement with a broad base, a movement that transcends traditional political boundaries. It is a movement that values people more than technology, people more than political boundaries and political ideologies, people more than profit."[93] Perhaps more than any national environmental organization, Environmental Action was about people. From its inception, Environmental Action operated with an organizational model not often found in the public interest sector at the national level. According the former Executive Director Ruth Caplan, "we function as a single collective."[94] With a staff of less than twenty employees and a board of less than ten members, collective decision making involved all members of the Environmental Action staff along with its board of directors.

Its last executive director, Margaret Morgan-Hubbard, sought to uphold the collective tradition at Environmental Action:

> History has shown that cumbersome bureaucracies and arbitrary hierarchies are unmistakable ingredients of the world's problems. We know that these dual evils, which rob people of their initiative and destroy their creative spirit, cannot be structures that contain meaningful solutions. Unlike some of our sister and brother organizations in Washing-

ton, D.C., EA is not compartmentalized with middle managers and division heads. We do not have hundreds of thousands of members, multitudes of staff or a multi-million dollar budget. We do not accept money from the large multinational corporations we seek to reform. We do not produce fancy high-tech, multi-media videos and slick advertising copy. Rather, EA specializes in answering phone calls and responding to member inquiries. We conduct our business in a democratic work place where each staff member's opinion is valued and taken into account. In turn, we work to facilitate the development of public policy that is fundamentally democratic both in its processes and in its consequences.[95]

Environmental Action retained its collective decision-making processes throughout its organizational restructuring. At the end, the organization bore little resemblance to its earliest structures, yet the management culture remained unchanged. Originally, Environmental Action was comprised of a 501(c)(4) parent organization, a 501(c)(3) tax-exempt foundation, and a political action committee, ENACTPAC. After its reorganization in 1994, the organization operated as a single 501(c)(3) tax-exempt entity. While the internal agenda-setting process continued to be collectively driven by all employees, the role of Environmental Action as a "small, national, activist, progressive organization within the ever-growing, complex environmental movement" did change as a result of its restructuring.

Environmental Action long maintained a reputation as "one of most dynamic, tenacious environmental groups on the national scene." Its "Dirty Dozen" hit lists of antienvironmental members of Congress, its hard-hitting critiques of federal policies, and its activist predisposition made Environmental Action a political force well beyond its size and wealth. Its tenacity was, in part, facilitated by its tax status. Unlike the vast majority of national environmental organizations, Environmental Action operated as a 501(c)(4) organization. As such, it faced no limitations on its lobbying or electioneering activities. It could use all of its limited organizational resources to advocate specific policies in Congress. Similarly, it could directly support or oppose members of Congress seeking reelection through its "Dirty Dozen" targeting or through its PAC contributions.[96]

After 1994, as a result of its new 501(c)(3) tax status, Environmental Action was limited in its ability to lobby the federal government directly, was prohibited from operating a PAC, and was severely limited in its electioneering efforts. Its "Dirty Dozen" hit list violated IRS electioneering standards for 501(c)(3) organizations.[97] By virtue of its restructuring, Environmental Action lost some of its organizational tenacity.

Beyond the restructuring of the organization, Morgan-Hubbard and the EA staff continued to wrestle with the limitations of their size and budgets resulting from their collective management philosophy. During its first year, Environmental Action attracted 10,000 members; in the intervening twenty-five years, its membership only doubled. In many ways, the staffers were faced with a technological catch-22. They needed money to implement their policies; in order to raise money, they needed the necessary fund-raising technology, but the necessary direct-mail technology was very costly, especially for smaller organizations.

Rose Audette, a former editor of *Environmental Action* magazine, expressed the concern of the entire organization. "A lot of it is bottom-line money. It takes capital to get members. It takes large amounts of money, and we're just not in that position. We don't have the amount of money it takes to double or triple our size. It's not feasible."[98] In the early 1980s Environmental Action made the decision to fire one of its lobbyists in order to purchase a computer so that it might attempt to maintain its existing membership at the very least. In the final years the organization subsisted by hanging on to its small but loyal following. Audette identified a block of charter members that remain fervent supporters of this liberal cause.

> There is a core of Environmental Action members of some unknown number who have been around since 1970 . . . and they love us to pieces, and they believe in us like perhaps we don't believe in ourselves. We see ourselves as being one of the most liberal, most left on the spectrum of the environmental groups, and we don't make any bones about that. Not all our members agree with us, but I would guess that most them do because this is the way we portray ourselves.[99]

Former Executive Director Ruth Caplan echoed the sentiment: "Our core membership wants us to get at the heart of the difficult environmental issues," but questions whether this strategy will allow the organization to endure. "Whether, in the 1980s and beyond, this is the basis on which to build a large membership organization is not yet at all adequately tested by our organization."[100] Upon taking over the organization in 1993, Margaret Morgan-Hubbard acknowledged the realities of organizational competition in the contemporary national environmental movement, and argues that Environmental Action will remain "small, lean and flexible. We have endeavored to be faithful to the notion that we must practice what we preach."[101] In many ways the demise of Environmental Action is an object lesson for the national environmental movement as well as for the larger public interest sector. The realities of organizational governance at the national level finally caught up with Environmental Action. Philosophical purity or the unwillingness to compromise coupled with collective maintenance strategies contributed to the demise of the organization. Environmental reporter Naftali Bendavid concludes that Environmental Action "failed in basic organizational skills. But the group's demise seems to say something about today's Washington. A creature of the 1960s, EA may ultimately have been too pure for today's politics. Environmental Action helped save a great many things—oceans, rivers, forests—but in the end it could not save itself. Simply put, EA ran out of money . . . EA never banished its inner hippie."[102]

At the end Environmental Action leaders did attempt several recruitment and fund-raising efforts but either ruled them out or lost money as a result of implementing the campaigns. In the early 1990s a direct-mail campaign was aborted as initial prospecting yielded little gains at potentially significant costs associated with a full-scale mail campaign. In 1995 the organization attempted a door-to-door canvassing campaign but also lost money even while gaining new members. In June 1996 the decision to close down operations was made. Environmental Action officially ceased operations in October of that year.

A number of Environmental Action staffers blamed executive director Margaret Morgan-Hubbard for her poor management during her three-year tenure. According to former EA attorney David Monsma, now working at the Center for Economic Priorities, "Frankly the organization needed a good manager, and I am not always sure we had the best management."[103] Morgan-Hubbard was also criticized for changing the policy agenda by adopting campaign finance reform as a core issue for Environmental Action.

Whether due to leadership weaknesses or ideological purity, it is indeed unfortunate when the national environmental movement loses one of its most feisty, combative, and intransigent elements to the organizational realities of the late 1990s.

The Leadership Challenge for the Next Century

The analyses of these five organizations highlight the difficulties of environmental leadership at the national level. On a daily basis, organizational leaders are faced with often conflicting priorities relating to organizational maintenance and policy advocacy. Critics of the national component of the environmental movement have offered a wide array of reasons for the difficulties that face many of the nation's major environmental organizations.

In a December 1994 study, "Restructuring Environmental Big Business," conducted by Christopher Boerner and Jennifer Chilton Kallery for the Washington University Center for the Study of American Business, the financial records of the Natural Resources Defense Council, National Audubon Society, Sierra Club, The Wilderness Society, and the Nature Conservancy were analyzed. The authors of the report identify mismanagement as one of the key sources of decline in the environmental movement: "By significantly expanding the scale and scope of their groups' activities, the directors of environmental organizations hoped to capitalize on Americans' increasing demands for environmental quality. Unfortunately, as many U.S. corporations have discovered, expansion away from an organization's core competency often has numerous disadvantages."[104] With increasing revenues in the 1980s many organizations expanded their research and advocacy agendas to include "more scientifically complex and politically difficult issues such as global warming and saving obscure animals. To support such work, the groups expanded their lobbying and public relations staffs, particularly in Washington, and helped to pioneer the media-driven techniques of modern public interest campaigns."[105] As the economy began to sour in the early 1990s, organizations were forced into a fierce competition for financing. In order to "differentiate themselves and rally public support, environmental groups began employing highly emotional, often misleading, campaigns," the report stated. "Organizations have used fear and apocalyptic prophecies to promote their political objectives." The report concludes that such tactics helped generate a powerful countermovement from within the scientific community, the business community, and also from state and local governmental officials and private citizens.

Several environmental leaders have questioned the validity of the study. John Adams of the Natural Resources Defense Council dismissed the report as "a political document and not a scientific study."[106] Adams argued that the Center for the Study of American Business, directed by Murray Weidenbaum, former chairman of President Reagan's Council of Economic Advisers, reflects the same antienvironmental views held by the Reagan administration. Nonetheless, other leaders in the environmental movement have recognized in their own organizations the problems outlined above. Barbara Dudley, then executive director of Greenpeace, U.S.A., in attempting to reorganize her organization acknowledged the problems associated with the rapid growth of the 1980s: "The national groups, including Greenpeace, did become too large in the 1980s, and did grow top-heavy and too focused on legislative solutions to environmental problems. As with many social movements of the last half century—the labor movement, the civil rights movement, the women's movement—the draw to legislative solutions was too seductive, and took the organizations too far from their grassroots."[107]

Political sources on Capitol Hill in Washington echo the sentiments of Dudley. A Democratic environmental committee staffer argues that "they've become too corporate, too 'Washington.' They've concentrated on building their [membership] lists and having their lobbyists and having their buildings and being insiders." Another source who works closely with many national environmental leaders concludes: "They're into their marketing and their direct mail, but they have totally lost touch with and offended the grassroots groups."[108]

Grassroots activists within the environmental movement also view the changes that have occurred at the national level with disdain. Karyn Strickler, formerly director of the National Endangered Species Coalition, found her own attempts to link the national organizations with grassroots groups thwarted by national leaders.

> Many national environmental leaders are stereotypical "ivory tower elitists." They seemed to feel that speaking in abstruse language conferred upon them a superior status, and they refused to make a case ordinary Americans could rally around. Moreover, each group believed it needed its own message and activities to distinguish itself from other groups, supposedly competitors. During my tenure, the national environmental groups gave no more than lip service to grassroots involvement in the [endangered species] reauthorization campaign. . . . National leaders turned a deaf ear to grassroots activists themselves. I was fired from my job for communicating my thoughts about the timing of the reauthorization with thirty grassroots groups, asking them to share their thoughts with the national environmental leaders. A senior vice president at the National Audubon Society said, "How dare you lobby me. I don't need to hear from the grassroots. I know what the grassroots thinks." . . . The kind of world my son inherits may well depend on the ability of national environmental organizations to recognize their problems and radically reconstruct themselves.[109]

Organizationally, the national environmental movement has become "top-heavy" and "too 'Washington.'" Throughout the last fifteen years, while the membership and development staffs of the various national organizations were competing for resources, there has existed a consistent level of interaction and support among the policy staffs of these groups. From the formation of the Group of Ten to the more contemporary Green Group, national environmental organizations have joined in coalitions to support or oppose a wide variety of federal governmental activities.[110] Unfortunately these interactions were too often conducted by the leadership staffs in Washington and New York and have failed to integrate the grassroots components of the organizations.

Mark Dowie, in his recent book, *Losing Ground: American Environmentalism at the Close of the Twentieth Century*, calls for a fourth wave of environmental activism found at the grassroots, "angry and impolite," and willing to "advocate nonviolent militancy to remind politicians at every level that the majority of their constituents consider environmental protection a government responsibility."[111] While Dowie has identified the missing element in the contemporary national environmental movement, his fourth wave of grassroots activism that jettisons the national organizational structures is destined to produce a loud, yet unintelligible cacophony of voices from the grassroots.

The leadership challenge for next century is to harness the energies of grassroots activists and to incorporate them into a coherent strategy of advocacy. A new wave of grassroots activism by the thousands of state and locals environmental groups, in the absence of organizational leadership at the national level, will be lacking in the "glue that holds them together or makes them a force."[112]

In order to meet this challenge, national organizational leaders must attract and retain a viable membership base. In the next chapter the methods of membership recruitment and retention used by national environmental organizations today—direct mail, telemarketing, and canvassing—will be analyzed, and their ability to maintain memberships in an effective and efficient manner will be assessed.

4

Membership Recruitment and Retention
Direct Mail, Telemarketing, and Canvassing

Our business is basically communications and we've lost the imagination of the average guy on the street.

—TED EUBANKS, MEMBERSHIP DIRECTOR,
NATIONAL AUDUBON SOCIETY

The only option is to go back to the people and talk to them, to try to find out why these issues don't seem to resonate—to try to find the embers and fan them into life once again.

—BRUCE BABBITT, SECRETARY OF THE INTERIOR

National environmental organizations would cease to exist without the support of individual contributors.[1] As a result, the recruitment and retention of members by the leaders of these organizations are matters of great importance. Significant staff time and organizational resources are dedicated to the nurturing of membership. Currently there are three main methods of membership recruitment and retention used by large public interest membership organizations: direct mail, telemarketing, and canvassing.[2] Each method has its strengths and weaknesses, but when used in combination, together with the proper messages, these techniques have produced positive results for a wide variety of organizations.

The Check's in the Mail: Direct-Mail Marketing

Direct-mail marketing in the United States is big business. According to the most recent data available from the Direct Marketing Association, the umbrella organization for the direct-mail industry, the United States Postal

Service delivered just under 70 billion pieces of direct mail in 1995, accounting for more than 40 percent of the total volume of mail processed. Since 1980 the volume of direct mail delivered in the United States has doubled. Total annual revenues generated from direct-mail solicitations, catalogs, and advertising will likely surpass $400 billion before the end of the century.[3] Of course, not everyone looks upon direct mail as an important and necessary component of American life. In fact, direct mail is more commonly viewed as wasteful "junk mail" by many consumers. "According to Earth Works Group, two million tons—tons, not pounds—of junk mail are sent to Americans each year. That's about a tree and a half for every adult. Fully 44 percent of all junk mail is never opened."[4]

Most direct mail is delivered to our homes at third-class mail rates. The United States Postal Service offers three rates at which third-class mail is charged: a basic rate for unsorted mail, a 3/5 digit presorted rate, and a lowest rate for direct mail sorted by postal carrier routes. The same three-tier rate system is offered to nonprofit organizations, but at lower rates. Those who have received some quantity of direct mail from nonprofits quickly recognize the distinctive nonprofit stamps attached to membership solicitations or renewal letters. Under current nonprofit mailing rates, a large nonprofit organization with an elaborate internal printing and sorting system with high-volume mailing is able to deliver a membership solicitation to your home for less than a nickel, not the 33 cents we pay at the post office. Without such rates, though, direct-mail recruiting for nonprofits would never be cost effective.

Nonprofit direct mail accounts for approximately 20 percent of the total direct mail volume, or roughly 14 billion pieces annually. Nonprofit mailers include trade, industry, and professional associations, unions, churches, veterans organizations, educational institutions, charities, museums, and political organizations such as national environmental organizations and the American Association of Retired Persons.[5]

The application of direct-mail marketing to public interest organizational maintenance during the last thirty years has dramatically changed the public interest sector, both quantitatively and qualitatively. "Using direct mail to generate donations to environmental groups was pioneered in the 1950s by the Defenders of Wildlife and the National Parks and Conservation Association. It was adopted wholeheartedly by the National Audubon Society in the 1960s and today is ubiquitous and, apparently, essential."[6] Beyond the environmental movement, Common Cause popularized this mobilizing technology; its efforts in the early 1970s were quickly imitated by many other organizations.[7] The conservative wing of the public interest sector has also made use of this technique. Writing in 1980,

Richard Viguerie, guru of the New Right, was convinced that "the conservative movement is where it is today because of direct mail. Without direct mail, there would be no effective counterforce to liberalism, and certainly there would be no New Right."[8]

The impact of direct mail as a recruiting device as well as a method of retaining existing memberships is evident in the national environmental movement. Table 2.7 indicates how rapidly four of the five organizations under discussion have grown since the mid-1970s when direct mail as a way to maintain membership was implemented on a large scale. With the exception of Environmental Action, which could not afford to use the technique for recruitment purposes, these environmental organizations have had dramatic increases in memberships since 1975, and direct mail is one of the main reasons for this growth.

Direct mail is treated very much as a scientific enterprise. All of the accumulated skills of Madison Avenue consumer marketing are now a part of the public interest arsenal of maintenance strategies. Direct-mail methodologies are continually tested and refined. The underlying strategy of direct mail, however, is quite basic: create a message, mail the message to citizens, and wait for a response. Implementing the basic strategy is more difficult. The technology of direct mail is so advanced, however, that it is virtually impossible to fail if one follows all of the rules.

Given that Ken Godwin has published an entire book on the topic of direct mail, *One Billion Dollars of Influence*, the implementation process is only briefly outlined here.[9] The implementation effort begins with the message. According to Roger Craver, one of the top direct-mail specialists for liberal causes and founding partner in the consulting firm of Craver, Matthews, Smith, and Co., argues that direct mail is particularly suited to public interest organizations: "Direct mail is a medium of passion, when used politically. Candidates and causes which can easily arouse that passion are naturals. But, whatever the issue, a direct mail letter must evoke a strong emotional response to be successful."[10] Organizations competing for members must make a pitch to potential contributors that motivates them to respond. Listed below are some examples of the direct mail that emphasize specific concerns; each letter begins with the salutation, "Dear Friend."

> As a United States citizen you have the right to vote. And while it's not written into the Constitution, you have the right to safe drinking water. You are also entitled to breathe unpolluted air, and you are certainly entitled to eat wholesome food uncontaminated by chemical poisons.
>
> *(League of Conservation Voters)*

TABLE 4.1 Direct Mail: Content Analysis, Part 1

ORGANIZATION*	NO. OF PIECES	OTHER VERSIONS	DECAL	PREMIUMS	PHOTO.	PETITION	NO. OF ITEMS
World Wildlife Fund	11	3	Y	4,8	2	N	5
Greenpeace	9	4	Y	6	1	Y	6
Common Cause	7	5	N	N	1	Y	7
Union of Concerned Scientists	5	4	N	N	1	N	6
NRDC	5	2	Y	N	1	N	5
Defenders of Wildlife	4	3	Y	4,7	1	N	5
UNICEF	4	2	N	N	1	N	5
Handgun Control	4	2	N	N	1	Y	4
MADD	3	2	Y	10	1	N	3
National Audubon Society	3	1	N	1,2,3,4	2	N	6
Amnesty International	3	1	Y	N	1	N	4
Gorilla Foundation	3	1	N	5,6	2	N	4
The Cousteau Society	3	1	Y	N	1	Y	4
Nature Conservancy	3	2	Y	9	N	N	5
ACLU	3	1	N	N	N	Y	4

*These organizations are selected from the mail received by the author during an eighteen-month period. Y = Yes; N = No; Other Premiums: 1 = bird feeder, 2 = backpack, 3 = free trial membership, 4 = pin, 5 = poster, 6 = stamps, 7 = photography, 8 = calendar, 9 = membership card, 10 = address labels. Photography: 1 = black and white, 2 = color; Photo = photography presented in mailing; No. of Items = number of items in each mailing.

Ironically, the "last of life" for millions of Americans is far from the best. Retirement incomes seldom keep pace with the rising cost of survival.

(Gray Panthers Project Fund)

Certainly the most troubling scandal to hit our federal government since Watergate is the escalation of money being poured into our nation's congressional elections by political action committees—the PACs.

(Common Cause)

Do the National Forests exist to serve the needs of the timber industry? Does the Forest Service? Outrageous questions, but to see and hear the industry in action you'd think the answer to both is yes.

(Sierra Club Legal Defense Fund)

The content of direct mail is determined through a painstaking process of drafting letters, copy editing, testing, revising, and packaging of the message. The passages quoted above were not written haphazardly. In fact, they probably were not written by the organizations at all, but by professional direct-mail specialists like Craver or Viguerie. These specialists have tested numerous versions of direct-mail packages to determine the most effective means of mobilization. They have even identified the key words and phrases that generate the highest response rates. The direct-mail industry in this country is large enough that several monthly trade magazines cater exclusively to the needs of that sector.[11]

Through a content analysis of more than 150 pieces of direct mail that I have received, several generalizations may be made. Tables 4.1 and 4.2 include several of the characteristics of direct mail. Of the mail received, World Wildlife Fund was the most frequent contributor, with eleven pieces sent in less than two years. These tables include information about the content of the more frequent mailers in the public interest sector (N = 70 pieces). The majority of the organizations offer some sort of premium for joining. For example, four of the five environmental organizations examined here offer "affinity" Visa or Mastercard credit cards as incentives for joining.[12] Decals are offered regularly as premiums as well. Petitions to be signed or surveys to be filled out are also frequently used methods for mobilizing new members. While the surveys lack validity or reliability as measures of the membership's preferences, they do serve to link the respondents to the organization and, hence, are successful recruiting tools. Most, but not all, organizations make some mention of their general goals or the important issues they are attempting to address. One major excep-

TABLE 4.2 Direct Mail: Content Analysis, Part 2

Organization	Rush Return	Organization Goals	Specific Issue Appeal	Member Benefits
World Wildlife Fund	Y	Y	N	N
Greenpeace	N	Y	Y	Y
Common Cause	Y	Y	N	N
Union of Concerned Scientists	Y	Y	Y	N
NRDC	N	Y	Y	Y
Defenders of Wildlife	Y	Y	N	Y
UNICEF	N	Y	Y	N
Handgun Control	N	Y	Y	N
MADD	N	Y	N	N
National Audubon Society	Y	Y	N	Y
Amnesty International	N	Y	Y	Y
Gorilla Foundation	N	Y	N	Y
The Cousteau Society	N	Y	N	Y
Nature Conservancy	N	Y	Y	N
ACLU	Y	N	Y	N

tion to the trend is the National Wildlife Federation. The initial proselytizing literature that I received contained no mention of the Federation's policy goals or issue agenda.[13]

The physical composition of the direct-mail pieces is also guided by direct marketing research. Researchers, for example, have found that people are most likely to read a four-page, single-spaced letter over any other length, shorter or longer. People also respond in greater numbers to envelopes with "live" first-class stamps rather than metered third-class postage. Envelopes range from white, business-size ones to the sophisticated packages offered by Common Cause and the American Civil Liberties Union and Handgun Control, Inc. The "People Against PACS" campaign conducted several years ago by Common Cause took the form of a file-folder, filled with petitions, press clippings, a letter from the executive director, and, of course, a response card for joining the organization.

The ACLU has packaged its mailings in the form of a legal document that is filled with indictments against various public officials, for example, Edwin Meese, when he was attorney general in the Reagan administration.

Finally, a mailing from Handgun Control, Inc., wins the award for the largest piece of direct mail; the package HCI leaders mailed out measured 13″ by 17″.[14] Even with government-subsidized postage, these larger pieces of mail are rather costly to produce and disseminate, which leads to the issue of making direct mail a profitable enterprise for public interest organizations.

The most terrific piece of direct mail is worthless until it is placed in the hands of individuals who will respond to the call for support. With the proliferation of organizations in the public interest sector, this is becoming an increasingly difficult task. The larger environmental organizations in the industry are mailing out between 30 and 60 million pieces of mail a year, and some mail more than 200 million pieces annually. The problem with such massive mailing efforts is the saturation of one particular attentive constituency.

This saturation is exacerbated by list trading or renting, a common occurrence in the public interest sector, especially within the environmental community. Many national environmental organizations have generated significant revenues by renting their membership lists to nonprofit and for-profit entities (see table 4.3).[15] The more innovative organizations are purchasing targeted mailing lists to "prospect" for new members. Environmental organizations are buying L.L. Bean and R.E.I. catalog lists, Diners Club membership lists, magazine subscription lists, including *Harpers, Atlantic, Mother Jones, Esquire,* and *Country Journal* as well as more targeted

---

TABLE 4.3 Mailing List Rental Income

Environmental Organizations	Income
Environmental Defense Fund	$181,665
Greenpeace	559,486
National Audubon Society	772,738
National Parks and Conservation Assoc.	98,586
National Wildlife Federation	680,626
Sierra Club	N/A*
The Wilderness Society	213,263
World Wildlife Fund	726,051

Source: Peter Overberg and Linda Kanamine, "Green But Not Growing," *USA Today,* October 14, 1994, p. 8A.

*The Sierra Club also generates substantial income from list rentals; unfortunately there is no line item in its annual financial statements that identifies list rental income.

environmental publications such as *E, Garbage,* and *Buzzworm.* They are even purchasing donor lists from art museums, zoos, theaters, opera companies, symphonies, and philanthropic organizations in efforts to expand the existing market.

When I began my research a decade ago, membership lists were purchased from professional list brokers at a rate of $55 per 1,000 names. Today, the cost has risen to $75 per 1,000 names. While the list rates have not increased significantly, the costs related to the production of direct mail and the associated mailing costs have driven the price of direct mail to more than $300 per 1,000 pieces. Obviously, when organizations are mailing out tens of millions of pieces annually, direct mail becomes an expensive endeavor.

Fortunately for the organizations, most direct mail campaigns require only a 1–2 percent response rate to break even. Most environmental organizations manage to recoup their startup costs with their initial mailings. Some even do extraordinarily well with certain campaigns. *Direct Marketing* magazine provides awards for mail campaigns that generate high response rates. Greenpeace, for example, received an Echo Award a decade ago from the magazine for its *Rainbow Warrior* mailing, following the bombing of its original ship in France. The package included a blueprint of its new ship, the *Rainbow Warrior II,* then under construction, that was "suitable for framing." A single mailing generated a 9 percent response rate, adding 43,000 new members to the organization.[16]

Direct mail, at the same time, has several negative consequences. First, the saturation of an existing market has led to the "graying" of the environmental membership constituency in the United States. According to Richard Hammond, president of Names in the News, one of the largest list brokers for liberal causes in the country, "the giving community is approaching sixty." When asked how will the environmental industry continue to exist if this pattern continues, he first laughed and said, "I'll be retired by then, it's not my problem." More seriously, he is concerned with the graying pattern but offers little in terms of a solution.[17]

Ten years ago the development directors of major national environmental organizations gave little thought to the prospect of graying membership bases. At the time virtually all organizations were flush with cash from their existing recruitment efforts. The most common response I received regarding the search for new members who were significantly younger than the existing pool of contributors was that younger folks, particularly college students, are not cost effective to recruit. They never give beyond the lower student rate; they want services and programs that the older contributors would not use, and they are transient, too hard to track. As a result, as recently as "1988–1989, there was no environmental organization specifically geared to a student constituency."[18] Furthermore, none of the national organizations had established campus outreach programs.

Today, a dramatic change is occurring; the national environmental organizations are now recognizing that their futures depend on a new generation of environmentalists. The first organizations to respond to the student constituency in 1990 were the National Wildlife Federation, the Student Environmental Action Coalition (SEAC), and Earth Day 1990. Since that time many of the national organizations, including the Sierra Club, through Sierra Student Coalition, The Wilderness Society, and Greenpeace, have developed strong campus outreach programs.

In addition to SEAC in Chapel Hill, North Carolina, a number of student organizations have formed across the country, such as Campus Green Vote in Washington, D.C., Green Corps in Philadelphia, and the Progressive Student Network in Chicago. Following the first national student conference at the University of North Carolina in Chapel Hill in 1989, which attracted more than 2,000 students, successive student conferences have occurred at the University of Illinois in Urbana-Champaign, Yale University, and the University of Pennsylvania, each attracting thousands of students. In addition, scores of campus-linked environmental websites now serve students, faculty, and administrators in promoting environmental awareness.[19] The national environmental organizations, while late to recognize the strength and vitality of younger environmentalists, are taking the necessary steps to incorporate them into their organizations and link them into their direct-mail networks.

In addition to the problem of the graying of the environmental constituency, there is a regional bias to direct mail efforts, particularly within the environmental industry; the evidence is shown in table 4.4.[20] Environmental membership percentages include members of the ten largest environmental organizations, divided by region. It is clear that the South is significantly underrepresented in the national environmental movement. Existing direct-mail efforts will only continue to perpetuate this bias.

Overcoming the regional biases in existing direct-mail lists is a difficult task. The solution to this problem, however, may also be found in the campus outreach efforts. On a large number of college campuses across the South, local environmental organizations have formed in the last five years. Creative steps to link these local groups to the national organizations will yield significant dividends in the future.

The final problem associated with direct-mail marketing for national en-

TABLE 4.4 Political Geography of Environmental Organization Membership

	REGION			
	Northeast	Midwest	West	South
% Environmental Membership	28.9	26.7	24.4	20.0
% General Public	23.7	25.5	19.4	31.4

Sources: Environmental Membership: Kathleen Ferguson, "Toward a Geography of Environmentalism in the United States," (Master's Thesis, California State University, Hayward, 1985); General Public: United States Census Bureau, 1980 Census Report.

vironmental organizations is alluded to in the quotations opening this chapter. Simply understanding the methodology and mechanics of direct mail does not guarantee a successful recruitment effort—the message matters. According to Roger Craver, "the movement has lost support because it has failed to demonstrate convincingly that its goals are compatible with the economic needs of the country."[21] In the process of communicating with prospective members through direct mail solicitations, a significant proportion of the national organizations have failed to offer messages that are sufficiently salient to attract new members. Recall the changes in the political climate identified in chapter 2. The continued economic uneasiness that pervades the American public despite major economic indicators reflecting prosperity has shaped the environmental debate through the late 1990s. Expressions of the apocalyptic consequences of environmental degradation have not energized potential environmental organization members.

In addition to the constraints on substantive environmental policy messages that seem to work in direct-mail recruiting efforts, the membership development directors of all 501(c)(3) nonprofit organizations are now chilled by the 1996 Internal Revenue Service ruling regarding electioneering in fund-raising letters.[22] In a March 1996 ruling, the IRS found that a particular 501(c)(3) organization, identified only as "M," had violated the prohibition of electioneering by including in its fund-raising direct mail partisan attacks upon individual elected officials as well as the phrase, "Together, we can change the shape of American politics." The IRS stated that this phrase, combined with the letter's description of a particular race for federal office, constituted "voter direction," not "voter education."[23]

The organization "M" is believed to be an abortion rights group that was active in federal and state elections, including the 1989 gubernatorial races in Virginia and New Jersey. The use of names of members of Congress together with action statements related to electoral outcomes is certainly not unique in direct-mail fund-raising. "The IRS is shooting straight at the heart of a rather common practice," said University of Miami law professor Frances Hill. "They are saying to charities, 'We know you are all doing this. We still say it's wrong.'"[24]

Nonprofit tax attorney and former IRS official William Lehrfeld believes that the actions of the IRS are a clear departure from previous policies but should not be ignored. The very fact that the IRS even considered evaluating direct-mail letters as part of an audit shows that it is serious in its pursuit of violators. "Look, a fund-raising letter has always been a fundraising letter, a tool designed to separate a fool and his money. Despite everything we may have spent learning from the IRS in years gone by, a fund-raising letter can now have a substantial impact on a group's 501(c)(3) status."[25]

For the national leaders of 501(c)(3) environmental organizations, the implications of the IRS ruling are significant, particularly for their direct-mail efforts conducted during electoral campaign seasons. While these organizations now have readily identifiable enemies on Capitol Hill, it will definitely require attention, time, and effort on the part of the wordsmiths crafting the direct-mail recruiting letters for the organizations to remain in compliance with the newly expanded definition of electioneering.

More than a decade ago there were some leaders in the national environmental community who talked of abandoning direct mail as a recruiting tool. For example, Alden Meyer, past director of Environmental Action and the League of Conservation Voters, argued that "the universe of direct mail lists is very limited and it's overtapped by all groups."[26] Efforts of the past decade have proven him both right and wrong. There is significant saturation of existing direct-mail markets; nevertheless, virtually all national organizations use direct mail with varying degrees of success.

Today, national environmental membership development leaders are inextricably tied to direct mail marketing. The recruitment and retention cycles have addicted organizations to this method. It is virtually impossible to wean an organization from direct mail without killing it in the process.[27] For the foreseeable future, large national organizations will remain committed to direct-mail marketing. This does not mean that direct mail is the only means of recruiting and retaining memberships. There are two additional methods of solicitation used by a growing number of public interest organizations—telemarketing and canvassing.

Dialing for Dollars: Telemarketing

Like direct-mail marketing, telemarketing in the United States is a big business. In fact, total sales generated in 1995 through telemarketing were greater than those resulting from direct-mail marketing; more than $385 billion in services and merchandise were purchased over the telephone. The American Telemarketing Association predicts that by the year 2000 the annual revenues generated through telemarketing will reach $600 billion. In 1993 more than 68 million individuals purchased products or services through telemarketing. In that year more than 60 million toll-free calls were placed to the nation's 1.8 million 800-numbers.[28] Today, there are well over two million 800- and 888-numbers in use in the United States.

Every year more than 4 billion telephone solicitations are made to people across the country. Unlike direct-mail solicitations, telephone contacts are more intrusive and are more often fraudulent. "Everyday, 18 million Americans are subjected to telemarketing 'sales pitches,' the abuses of which can range from simple annoyance to devastating loss," states Daniel Borochoff, president of the American Institute of Philanthropy, a charity watchdog organization.[29]

In an effort to combat the increasing level of telemarketing fraud in the United States, Congress passed and President Clinton signed into law the Telemarketing and Consumer Fraud and Abuse Prevention Act in August 1994. The law empowered the Federal Trade Commission (FTC) to promulgate regulations concerning deceptive telemarketing practices. Regarding the use of telemarketing in the public interest sector, the FTC has no jurisdiction over nonprofit organizations; however, it does have the power to act upon professional telemarketing firms hired by nonprofits.[30]

Due to the rather negative reputation of telemarketing, its inherently intrusive nature, and its comparatively high cost, most national public interest organizations do not use telemarketing as a means of "prospecting" for new members. Rather, the organizations that use telemarketing have found it to be an effective method of renewing members. Kathy Swayze, a private telemarketing consultant with Herzog Swayze, Inc., works with a number of environmental organizations on their telemarketing efforts. She argues that prospecting is simply not effective for large national organizations due to the low response rates associated with calling individuals unfamiliar with the organization and the high costs per contact. To her knowledge, only Mothers Against Drunk Drivers (MADD) has successfully used national telemarketing to recruit new members. Swayze did mention that gay rights organizations have used telemarketing successfully, but only when calling numbers from lists gathered at rallies and protests.[31]

As a result, national environmental organizations most often use telemarketing to renew existing members or to solicit additional contributions from their current members.[32] In these cases, the telephone respondents are familiar with the organization and are less likely to reject the telemarketer at the outset. According to Swayze, organizations have two options when using telemarketing—contract out with a telemarketing firm or conduct the effort in-house using volunteers or staffers.

When contracting out, telemarketing firms offer two basic rate schedules. The first billing method involves paying by caller/hours while the other method involves billing for each contact made by the callers. In the first instance, a firm might charge $35–$40 per hour per caller for 8–12 contacts each hour. In the second instance, the charge of $4–$6 per contact would be assessed to the client organization, with high-dollar contacts costing as much as $15. In either case, telemarketing is an expensive enterprise. Obviously, in-house volunteer efforts are cost effective, but it is quite difficult to mobilize individuals for such efforts, particularly for national organizations without grassroots infrastructures.

In spite of the costs or difficulties associated with telemarketing, national organizations have found telemarketing to be a smart investment of time and resources. My own experiences have shown that it is much easier to throw away a renewal notice than it is to hang up on someone asking for a membership renewal, although I rarely succumb to giving additional contributions through telemarketing. Telemarketing, while serving well as a retention tool for national environmental organizations and generating additional funds from existing members, does not serve to expand the organizational base of support. Moreover, its inherently intrusive nature may make potential enemies out of former friends.

While these two techniques, direct-mail marketing and telemarketing, may attract slightly different audiences to the wide array of liberal and conservative public interest organizations soliciting members across the nation, these approaches are quite different from the social network mobilization efforts of the earlier environmental movement. As a result, the memberships being attracted to contemporary environmental organizations are qualitatively different from the members a generation ago. These electronically recruited individuals have some degree of loyalty to a particular cause but little loyalty to a particular organization.[33] Political theorist Sheldon Wolin concludes that the contemporary public interest movement has none of the characteristics of earlier social movements: "The new politics has special conceptions of membership, participation, and civic virtue: a member is anyone on a computerized mailing list, participation consists of signing a pledge to contribute money; civic virtue is actually writing the check."[34]

In many ways, canvassing, the other method of membership recruitment used by a number of national environmental organizations, has reintroduced the social network mobilization component as a means of membership recruitment and retention.

Knock, Knock, Knockin' on Donors' Doors: Canvassing

Unlike direct-mail marketing and telemarketing, canvassing is a method of sales and recruitment that has a history going back to the beginnings of commerce and politics in the United States. Despite the longevity of this method, canvassing or door-to-door marketing of consumer products and services or political candidates and causes also has its detractors. From the traveling "snake oil salesmen" of centuries past to the contemporary gypsy roof repairmen who scam the elderly, direct-to-home marketing has its fair share of fraud and abuse.[35] Nevertheless, there are several quite profitable companies in the United States that have based their entire marketing strategies on direct selling, such as Amway, Tupperware, Avon, Mary Kay Cosmetics, and Discovery Toys.[36]

From a sales perspective, door-to-door marketing in the United States generated $16.55 billion in revenues in 1994, according to the Direct Selling Association. Approximately half of the annual sales revenues are derived from personal care products (cosmetics, jewelry, skin care, vitamins, etc.), with an additional 25 percent from home/family care products (cleaning products, cookware, cutlery). The remaining 25 percent of annual sales include the marketing of services, subscriptions and memberships, leisure products (toys and games), and educational products (books, encyclopedias). These goods and services are marketed to individuals and groups by a national sales force of 6.3 million women and men.[37]

Canvassing efforts are not nearly as widely used as direct mail or telemarketing by national environmental organizations. There are a few organizations, however, that have used door-to-door canvassing with much success, Greenpeace, Sierra Club, and the League of Conservation Voters, for example. The difficulties of this method of recruitment lie in the relative cost ineffectiveness and concerns for safety of face-to-face contacting of citizens, particularly in the late 1990s.

I am old enough to remember a time when milk and bread were delivered directly to homes by "milkmen" and "breadmen." Today, such individualized delivery services are virtually nonexistent. We are becoming increasingly insulated from the outside world; our homes have become our fortresses. With gated communities, secured high-rise buildings, local laws prohibiting door-to-door solicitations, and an increasingly violent society, Americans are much less likely to open their doors to strangers. Even Boy Scouts and Girl Scouts have taken to hawking their wares at supermarkets and malls rather than going door-to-door, due in large part to concerns for the safety of the children. Apart from these societal changes, the most effective areas for canvassing are those in which there is a high population density. Apartment buildings, rarely accessible these days, are the most effective housing units to canvass; travel between units is a matter of several steps or a flight of stairs. Row housing neighborhoods also provide large numbers of closely clustered units. From a door-to-door travel perspective, canvassing in rural areas is not cost effective.

The environmental organizations that use canvassing have concentrated on urban and suburban locations throughout the country, such as Philadelphia and its suburbs and the Chicago area. The result of canvassing for organizations is a clustered membership base, useful in grassroots targeting of a few elected officials at a variety of levels of government, but the method is not as effective for broad grassroots mobilization at the federal level. For the most part, public interest organizations conduct canvassing on a voluntary basis. Paid canvassing efforts open the contracting organizations to liability in the cases of injury or harm to canvassers and are very expensive. However, contracting for canvassing services within the public interest community is not uncommon.

In 1993 the Sierra Club reintroduced canvassing into its membership development strategy. Contracting with employees of the Fund for Public Interest Research, an organization with ties to the Ralph Nader network, the Sierra Club canvassing project contacted more than two million people during its first three years of operation and successfully recruited more than 250,000 new members. Program coordinator Emily McFarland views canvassing as a crucial component in the revitalization of the environmental community nationwide: "This is one-on-one, face-to-face communication. Once you get your foot in the door, people will listen. And once they understand the threats to the environment, they'll take action."[38] Conversely, Environmental Action contracted out for canvassing services in 1995. While the recruitment effort yielded 5,000 new members, Environmental Action paid canvassers more than was generated by the new membership dues collected.

Despite its problems, canvassing does reintroduce the social network approach that created the environmental movement in the 1960s—neighbors and friends recruiting neighbors and friends at the grassroots level. Unfortunately, Americans nowadays are much less responsive to the knock on their door than they were thirty years ago. In theory, canvassing for members may result in greater attachments to the organization through the social networking process. In practice, membership recruitment and retention is a numbers game. Unless an organization is able to have literally thousands of volunteers canvassing on a full-time basis across the country or is willing to risk potential losses by contracting out for canvassing services, this method cannot produce the results of direct mail or telemarketing.

Additional Means of Organizational Support

For the vast majority of national environmental organizations, membership recruitment and retention involves the use of one or more of the aforementioned methods. There are several additional sources of external financial support that are not generated through massive solicitations but are nevertheless important in organizational maintenance. These institutional and individual patrons include foundations, corporations, the federal government, through grants and contracts, and large individual donors.[39]

Traditionally, foundation support has played a significant role in the formation of public interest organizations. Today, national environmental organizations continue to solicit foundation support for specific projects as well as for general organizational maintenance purposes.[40] The new Alliance for Environmental Innovation launched by the Environmental Defense Fund and the new Environmental Information Center each received significant funding from the Pew Charitable Trusts.[41] In addition to this support, EDF received contributions from more than 150 foundations, totaling more than $7 million in 1994. Even smaller organizations are able to attract the support of major foundations; Environmental Action, for example, received financial grants from more than twenty foundations in its last four years.

Recently a number of environmental interests joined forces in creating Earth Share, an umbrella foundation that solicits contributions in much the same fashion as the United Way and distributes its resources to forty-four national environmental organizations, including the Environmental Defense Fund, the National Wildlife Federation, the Sierra Club, and The Wilderness Society.[42]

Corporate financial support for national public interest organizations continues to be a source of conflict, particularly for environmental organizations. Greenwashing, or the buying of goodwill by corporations through contributions to environmental causes, is viewed by many environmentalists as an inappropriate intrusion by corporate America. Nonetheless, many environmental leaders are hard pressed to turn down these large corporate contributions. As discussed in the preceding chapter, the National Wildlife Federation has institutionalized corporate participation through its Corporate Council. Other national organizations, such as EDF, the Natural Resources Defense Council, the Nature Conservancy, the World Wildlife Fund, the National Audubon Society, the Izaak Walton League, and the Sierra Club, also receive significant corporate contributions.[43] In addition, several organizations have entered into licensing agreements with a variety of corporations. The Sierra Club, for example, has signed licensing agreements with more than a dozen companies to market products bearing the Sierra Club name, for which it received royalties of $1.6 million in 1994.[44]

The nonprofit tax status granted by the federal government is only one aspect of government support that facilitates the maintenance of public interest organizations. Specific government grants and contracts are also important elements in the support of national organizations.[45] For the national environmental movement, the Environmental Protection Agency and the Department of Interior as well as the Occupational Safety and Health Administration, the Department of Transportation, and the Department of Defense, among other executive agencies, provide competitive grants and contracts for which organizations are eligible. Jack Walker, in his 1985 analysis of more than 250 nonprofit organizations, found that almost 10 percent of the organizational budgets were derived from government sources.[46] Today, with federal government downsizing, both in personnel and budgets, government grants and contracts will be fewer in number, smaller in value, and increasingly competitive.

The final aspect of patron solicitation involves the cultivation of wealthy donors or "contributors of serious money."[47] Unlike the impersonal direct-mail and telemarketing strategies that generate hundreds of thousands of membership level contributions of $25 to perhaps $100, the solicitation of individuals willing to donate thousands of dollars through one-time contributions, annual giving programs, or estate bequests is an increasingly important component of patron support.

This type of solicitation involves personal networking at the elite level.[48] Individuals with disposable incomes sufficient to support large philanthropic contributions have come to expect a personal touch when being courted for contributions. In many cases these individuals are political and social insiders who routinely have direct contacts with political or organizational leadership, not impersonal contacts by telephone or through the mail. Brown, Powell, and Wilcox draw several distinctions between their "serious money" contributors and the mass-recruited contributors:

> Our data confirm the insider status of personal-acquaintance network contributors. Compared to those impersonally solicited, personal network givers are more likely to have held party office and public appointive office. They are also more likely to have been asked by others for political advice, to have been asked by others to help with the national government, and to have helped form a new group or organization to try to solve some community problem. Those solicited by mail or telemarketing are especially likely to have *written* to a policy maker, while personal-acquaintance network contributors are especially likely to have *spoken* with a policy maker. . . . Those contributors solicited through a personal network were more likely to be better educated, affluent, middle-aged, and male than those successfully contacted through impersonal networks.[49]

The characteristics of large individual contributors to public interest organizations would likely be quite similar to the presidential campaign contributors discussed above, perhaps even more affluent. As a result of these distinctions, virtually all of the large national organizations have staff in their membership and development offices specifically focused on the solicitation of large individual donors.

Not only has the competition for large contributors become more intense in recent years but there is also competition across the entire nonprofit sector for staffers adept at cultivating donor networks. In early 1996, for example, the director for major gifts at the National Wildlife Federation was hired away by Princeton University to serve as its major gifts coordinator. The solicitation process for serious money has also involved the policy staffs of some organizations, generating internal tensions. During Jon Roush's presidency of The Wilderness Society, regional staff members were called upon to serve as guides and escorts on wilderness trips for potential large contributors who were visiting their locations. A number of regional staffers balked at having to serve as tour guides for these wealthy contributors. Regardless of the internal tensions, however, the cultivation of large individual contributors will remain an important part of organizational fund-raising.

The Costs of Organizational Maintenance

National environmental organizations have come to rely on a variety of methods of membership recruitment and retention as well as the solicitation of institutional and individual patrons in order to maintain their organizations. For most people it is difficult to discern the degree to which the organizations they support are effective and efficient in their maintenance efforts. While nonprofit organizations are required by the Internal Revenue Service to file their 990 tax forms annually and are also required by many states to file their annual financial reports with the appropriate government agency when conducting direct mail or telemarketing in each state,[50] rarely do members know the details of organizational maintenance in their organizations.

In my earlier research on the issue of public disclosure of organizational information, I argued that the IRS and various state agencies were not sufficient resources for individual members to learn about the internal operations of their organizations and that it was necessary to create an autonomous nongovernmental entity that would serve as an independent judge of organizational effectiveness: "What may be needed in the future is an organization to monitor the public interest sector—Public Interest Watch."[51] Three years later, in 1992, Daniel Borochoff created the American Institute of Philanthropy (AIP) in St. Louis, Missouri. On a quarterly basis, AIP presents its *AIP Charity Rating Guide and Watchdog Report;* each issue of the report rates more than 250 nonprofit organizations in more than thirty substantive areas, such as abortion and family planning, animal protection, rights of the blind and visually impaired, civil rights and advocacy, consumer protection and legal aid, drug and alcohol abuse, environment, gun control, health and diseases, human rights, immigration, international relief and development, peace and international relations, rights of senior citizens, women's rights, and youth residential care.[52]

AIP rates nonprofit organizations on three main dimensions relating to the efficiency of organizational maintenance: cost to raise $100, percentage of budget spent on charitable purpose, and years of available assets. AIP considers $35 or less to raise $100 to be reasonable for most organizations; it views 60 percent or more in expenditures on the substantive purpose of the organization to be reasonable as well. In addition to these two dimensions, AIP also factors into its ratings the years of available assets, including cash and investment assets but not land, buildings, and equipment that is used by the organization in its operation. AIP views negatively those organizations with assets equal to three or more years of budgetary expenditures.[53]

TABLE 4-5 American Institute of Philanthropy Ratings: 1994, 1996

ENVIRONMENTAL ORGANIZATIONS	AVG. COST TO RAISE $100		AVG. % OF BUDGET SPENT ON CHARITABLE PURPOSE		AIP GRADE	
	1994	1996	1994	1996	1994	1996
Center for Marine Conservation	$28.50	$32.50	70.5%	71.0%	B–	C+
Defenders of Wildlife	30.50	37.50	69.5	59.5	C–	C–
Environmental Action	11.00	N/A	86.0	N/A	A	N/A
Environmental Defense Fund	25.00	26.50	71.5	71.5	B–	B–
Friends of the Earth	21.50	24.00	65.5	67.0	B–	B–
Greenpeace^a	35.00	44.00	63.0	51.0	C	D
National Audubon Society	32.00	34.00	70.0	68.0	B–	B–
National Parks and Conservation Society	26.00	26.00	71.0	71.0	B–	B–
National Wildlife Federation	35.00	36.00	62.0	69.0	C+	C+
Natural Resources Defense Council	19.50	16.50	68.0	71.5	C+	B–
Nature Conservancy	10.00	15.00	88.0	81.0	A	A
Sierra Club^b	18.00	37.00	73.0	66.0	B+	C+
The Wilderness Society	30.00	29.00	63.5	60.5	C–	D

Sources: AIP Charity Rating Guide and Watchdog Report (Summer/Fall 1994), pp. 7–8; *AIP Charity Rating Guide and Watchdog Report* (March 1996), pp. 7–8. None of the organizations presented held assets

^aGreenpeace Fund, Inc., a 501(c)(3) foundation affiliated with Greenpeace, received AIP grade of A– in 1996.
^bSierra Club Foundation and Sierra Club Legal Defense Fund were also graded by AIP in 1996, receiving an A– and B, respectively.

The organizational data used by AIP is collected from the nonprofit organizations as well as from the Internal Revenue Service and is analyzed on a continuing basis. There are two effects on the data using this methodology. First, there is a lag effect in analyzing organizational data. For example, the findings presented in table 4.5 for 1996 are based on organizational data from 1995 and earlier. Second, by constantly updating its files, AIP produces its ratings using multiyear data. As a result, the AIP findings should be viewed as trends rather than as individual year ratings.

The organizational maintenance efficiency ratings for the five environmental organizations analyzed in this work are presented in table 4.5, along with those of several other organizations for comparison purposes. Two of the three AIP dimensions are presented, as none of the organizations presented holds assets equal to two years of expenditures.[54] Ironically, of the five groups, Environmental Action fared best in the 1994 AIP ratings. By avoiding costly direct-mail marketing for new members, its fund-raising operation was cost effective, spending only $11 for each $100 increment. Without direct mail, however, Environmental Action remained an organization of less than 20,000 members. Each of the remaining four organizations highlighted in the previous chapter received passing grades, although The Wilderness Society slipped into the unsatisfactory range in 1996.

The Environmental Defense Fund shows a consistent pattern of maintenance, spending about $25 to raise $100, with almost three-quarters of its budget directed at substantive environmental concerns. The National Wildlife Federation spends a bit more than EDF on its fund-raising efforts, at $35 per $100 raised; yet while keeping its fund-raising costs fairly constant, it managed to increase its budgeting toward its educational and policy programs.

The Sierra Club, like Greenpeace, shows the effects of its difficulties with its direct-mail recruitment efforts in recent years. The Club's fund-raising costs more than doubled in the two-year period captured by AIP. Interestingly, the Sierra Club Foundation and the Legal Defense Fund (now Earth Justice LDF) were much more efficient in their separate fund-raising efforts, receiving grades of A- and B, respectively, in 1996. Finally, The Wilderness Society receives the lowest marks, due in part to its relatively costly fund-raising efforts and its comparatively low percentages of budgets focused on substantive issues.

Apart from a strong rating of the Nature Conservancy at the most wealthy extreme of the national environmental organizations, the vast majority of national organizations are operating satisfactorily. Nonetheless, there is clearly room for improvement in the efficiency and effectiveness of organizational maintenance of many national environmental organizations.

The contemporary national environmental movement persists as a result of adopting the various methods of mass membership recruitment and retention along with the solicitation methods that attract institutional and individual patrons. As a result, the movement bears little organizational resemblance to the conservation movement of thirty years ago. With the exceptions of the Sierra Club and the National Wildlife Federation and the newly created campus outreach programs in some organizations, there is little that is social about the movement, and the only movement one finds today is that of individuals from one mailing list to another. Given the realities of the environmental movement in the late 1990s, most environmental organizations will continue to exist, despite some significant problems in organizational leadership and management, as discussed in the preceding chapter. William Turnage, past president of The Wilderness Society aptly summarized the current state of maintenance among environmental organizations: "it's quite feasible to stay in existence, it is difficult to be effective." The contemporary environmental movement will endure, not because it is terribly effective as a collective policy advocate, but because of technological and organizational innovations that, for the time being, allow individual organizations to maintain sufficient membership bases and patron support to meet monthly payrolls.

The next two chapters explore the consequences of these organizational maintenance strategies by focusing more closely on the people that decide to join environmental organizations. These activists are analyzed in terms of their characteristics and attributes and their other political activities. Chapter 5 looks specifically at the motivational patterns of members along with the incentive structures provided by organization leaders. Chapter 6 then analyzes the linkages between leaders and members and evaluates the degree to which members, as activists and organization advocates, act in concert with the strategies employed by organization leaders.

5

It's Not Easy Being Green
Leadership Incentives and Membership Motivations

The prospect of a loss is more likely to motivate action than the
expectation of a gain even when the value of the gain and the loss are
identical. An environmental lobby will get more members by saying that
an existing national park faces a reduction in size by 10 percent than it
will by saying that the same park might be expanded by 10 percent.
—James Q. Wilson, *Political Organizations*, 1995.

Membership has its privileges.   —American Express

Think for a moment about what motivates you to belong to an organization—a church, a social group, a service organization, a sorority or fraterni-
ty, a professional association, or a public interest group such as an environmental organization. Is it the camaraderie enjoyed with friends? Is it the
mission of the group? Perhaps it is the tangible benefits provided by the
group, such as tote bags, calendars, glossy monthly magazines, or even discounts on travel and lodging or drugs and insurance. Or perhaps you are
motivated by fear or the threats of disaster if you do not act.

From an organizational leadership perspective, the keys to organizational maintenance in the public interest sector are attracting *and* retaining individuals like you and me as well as a sufficiently large membership and/or
donor base of support to sustain the daily operations of the organizations.
In order to attract and retain us, leaders must appeal to our motivations by
offering a combination of incentives that will appeal to a wide audience of
potential supporters. As demonstrated in chapter 4, direct mail, telemarketing, and canvassing are the methods of choice for conveying these incentives to potential members. In this chapter the interrelationships be-

tween the incentives offered by organization leaders in order to attract new members and retain existing ones as well as the motivations individuals have for belonging to these public interest organizations will be analyzed.

James Q. Wilson succinctly summarizes the broad theoretical underpinnings of the existing research on political incentives and organizational maintenance in a new introduction to his classic work, *Political Organizations*.[1] Acknowledging the purely economic transactions identified by Mancur Olson as defining the relationships between unions and their members, bar associations and lawyers, and medical societies and doctors based on the provision of direct economic benefits, such as collective bargaining for wages and benefits, legal certification, or reduced rates on malpractice insurance,[2] Wilson defines three additional areas of motivation that may provide incentives for potential and existing members of public interest organizations: (1) differences in temperament, (2) organizational learning, and (3) perception of threat.

Although national environmental organizations, and public interest organizations more generally, are largely defined in the political arena by the policy goals they pursue, these organizations are not precluded from marketing themselves by offering incentives similar to the economic benefits cited above to differentiate themselves in the public interest market. The organizational names and logos are themselves marketing tools. The recent name change from Sierra Club Legal Defense Fund to Earth Justice Legal Defense Fund is exemplary of the quest for product differentiation. In this instance, there was not enough market distance between the mix of Sierra Club "products," with multiple entities sharing the same Sierra Club label.

The vast majority of environmental organizations offers some form of selective incentives (i.e., for members only) to entice potential members into the organizations. These selective incentives appear in the form of premiums, such as tote bags, coffee mugs, affinity credit cards, calendars, T-shirts, posters, and decals; (see tables 4.1 and 4.2). For the largest organizations, such as the American Association of Retired Persons, such premiums include significant discounts on drugs and life insurance as well as reduced rates at major hotels and car rental agencies. For our purposes, the magazines and newsletters offered by organization leaders will also fall into this form of membership incentives. Wilson refers to these premiums as material incentives because they are tangible and have some market value independent of the organization. In the analysis that follows, we will identify this recruitment and retention dimension as *PREMIUMS*.

The next area of motivation identified by Wilson relates to the differences in temperament. According to economic rational choice theory, economically rational individuals should never take part in any activity when the costs of undertaking the activity outweigh the benefits. Voting, for example, would be an economically irrational activity as the costs associated with the act of voting (e.g., time, effort, expense) are higher than the benefits (i.e., the value of one vote in an election). Of course, tens of millions of citizens vote in the United States in each election. But tens of millions of citizens also stay home. To explain why individuals do vote or join public interest organizations that provide collective goods (e.g., consumer protection or environmental protection), Wilson offers two views on individual temperament. First, he argues that many people undertake certain activities, such as voting or joining public interest groups, based on a sense of civic duty—because "it is the right thing to do," regardless of personal benefit.

Terry Moe and other scholars have added to this dimension by focusing on the higher levels of "subjective efficacy" displayed by political activists[3] as well as the impacts of social pressures, feelings of responsibility, and ideological commitment.[4] Moe argues that some citizens in a society will overestimate the significance of their individual contributions to a particular collective action effort, leading them to join organizations from which they would gain benefits without joining, or through what Mancur Olson refers to as "free riding."[5] Implicitly linked to these two perspectives is a substantive component of the individual action. Wilson refers to this dimension as purposive because it relates to the substantive goals of the activity or organization that an individual wishes to pursue. In the analysis presented later in this chapter, this component will be labeled *GOALS*.[6]

The next area of motivation identified by Wilson is derived from the work of Lawrence Rothenberg and relates to the learning that takes place from the time an individual joins an organization until the time he or she leaves the group. Rothenberg's work on the behavior of the Common Cause membership has shed important light on the dynamic nature of membership motivations and the incentives that are used to recruit and retain them.[7] He posits that an individual may casually join an organization because the cost is low and the initial direct-mail literature was attractive. The individual, after ten years of membership, may remain in the organization for an entirely different set of reasons that may include social interactions as well as the policy goals of the organization. For organizations that offer opportunities for membership contacts through meetings, outing programs, national conventions, and travel or tour services—what Wilson calls solidary incentives—this dimension is an important element in membership recruitment and retention, despite its organizational costs. This dimension will be labeled *INTERACTIONS* in the analysis that follows.

The final area of motivation that may cause individuals to take political action is the perception of threat. The quotation by Wilson that opens this chapter captures the sentiment of this dimension. The prospect of a loss will motivate individuals to a greater extent than the prospect of a gain, particularly when the appeal for support is emotionally charged. As discussed in chapter 2, the perceived threat posed by the Reagan administration in the persons of Anne Gorsuch Burford and James Watt resulted in significant membership increases for virtually all national environmental organizations. Mark Hansen has found similar evidence of the positive effect of perceived threat on the memberships of the American Farm Bureau Federation, the League of Women Voters, and the National Association of Home Builders.[8] This dimension will be labeled *THREATS*.

The goal of this chapter is to provide a broader understanding of the mix of organizational incentives and membership motivations as they occur in real social and political contexts. In order to meet this goal, survey responses from members of the five environmental and conservation organizations presented in the preceding chapters will be analyzed: Environmental Action, Environmental Defense Fund, National Wildlife Federation, Sierra Club, and the Wilderness Society.[9] Although the surveys of more than 3,000 environmental organization members were conducted two decades ago, the motivations for belonging to public interest organizations and the incentives offered by organization leaders have changed little during the intervening twenty years. In fact, the responses to motivational questions by these group members have a richness that perhaps would not be found in survey responses today, due to the fact that there is a significant minority of respondents who joined these five organizations through social networks rather than through direct mail or telemarketing. Today, fewer members are attracted to these groups through social networks, due to the emphasis on high-technology membership recruitment.

Environmental Organizations and Membership Incentive Theories

Contrary to popular opinion that characterizes the "Green Lobby" as a monolithic force in American politics, all environmental groups are not alike; neither are environmentalists nor are their motivations for belonging to such groups.[10] While these five organizations are generally directed toward the collective goals of environmental protection and conservation, each group provides a different mix of incentives to maintain its membership. Beyond the research presented above, the theoretical work on organizational incentives may be found in several fields of inquiry including: economics (rational choice), political science (group theories as well as formalized rational choice), sociology (normative conformity), and psychology (affective bonding). As stated earlier, Olsonian rational choice models offer only selective incentives and public goods as explanations for the involvement of individuals in interest groups. The sociological literature incorporates the influence of societal norms as explanatory factors. Social action theory, particularly as presented in the work of Talcott Parsons, analyzes individual behavior as primarily normatively oriented.[11] Particularly when analyzing members of voluntary associations seeking collective goods, it is necessary to include a component that allows individuals to internalize, in a sense, public goods. The normative component is distinct from more concrete, goal-oriented, group-specific incentives. It includes the individuals' sense of responsibility and moral obligation coupled with perceptions of threat, or what Robert Mitchell has labeled "collective bads."[12]

In addition to rational choice and normative conformity as models of individual behavior, social psychologists have concentrated on the emotional motives for individual commitment to organized groups.[13] Affective bonding focuses on individuals' perceptions of their roles vis-à-vis other members and the organizational leadership. In some cases, affective bonding may result from purely symbolic representations—a sense of "oneness" based on some degree of affinity with a group's collective goals or perhaps its charismatic leaders. Bonding may also be related to interpersonal relationships formed as a result of group membership and/or leadership interaction.

Each of these motivational dimensions is required to offer a realistic account of individual motivations for belonging to public interest groups. As Knoke and Wright-Isak note, "Each implies a different 'logic' by which individuals reach decisions to contribute or to withhold personal resources from collective actions."[14] The rational choice decision-making process involves preference ordering among competing interests (i.e., favoring Group A over Group B and Group B over Group C). Normative conformity operates in a different sphere as actions are guided by individually unique and socially instilled values about appropriate behavior. Affective bonding focuses on the linkages or interpersonal identifications between individuals and/or symbolic representations in and of the groups. Together these theoretical perspectives are incorporated into the four motivational dimensions outlined above: *PREMIUMS, GOALS, INTERACTIONS,* and *THREATS.*

By synthesizing the theoretical propositions of Olson, Wilson, Parsons, Knoke, Rothenberg, Hansen, and Moe, a more inclusive incentive-motivation system may be constructed. The first component of the incentive system focuses on the tangible (utilitarian) incentives that motivate individuals to belong to public interest groups—premiums. Olson contends that only the smallest of organizations can exist without selective incentives or premiums, particularly those providing collective goods, due to the "free rider" problem.[15] Jeffrey Berry's analysis of public interest organizations showed that few groups offered substantial premiums yet still managed to maintain themselves. Three-quarters of the groups he studied did offer some form of publication but few went beyond that level.[16]

Within the environmental movement, however, groups' provisions of premiums range from a quarterly newsletter to very attractive, full-color magazines, outing programs, book discounts, clothing discounts, local group activities, and even a free Visa card. In view of the diversity of these incentives, it is plausible to assume that such incentives or the lack thereof will have unequal effects on individuals' motivations. An organization offering no premiums is very unlikely to attract members for this reason. However, it cannot be assumed that the presence of such incentives is either necessary or sufficient to motivate individuals to support a particular group. Nonetheless, a full incentive model would be incomplete without this component.

The next aspect of the incentive system incorporates the emotional, interpersonal, and symbolic attachments individuals may have to public interest organizations and their members. These attachments may be manifest in the social networks developed between members (and leaders) through facilitating organizational structures, group solidarity—a sense of "oneness" generated by group slogans, symbols, and other symbolic representations, and heightened personal status based on connections with influential persons or associations with charismatic leaders. One's commitment to an organization may have an affective component, that is, an emotional rather than a normative element. These manifestations are subsumed under the component labeled *INTERACTIONS*.

Moe argues that the substantive policy goals of interest groups are, in and of themselves, meaningful incentives and that individuals can and do belong to interest groups for various combinations of reasons.[17] For the purposes of this study, a distinction will be made between group-specific goals as motivations for belonging to a particular environmental organization, labeled *GOALS,* and the more general sense of responsibility and moral obligation generated by perceived threats, both individual and societal, that motivate members to belong to groups seeking collective goods and that are labeled *THREATS.*

THREATS involve individuals' recognition of the existence of a societal inequality or dilemma (in this case, threats to the environment) and their sense of efficacy, which causes them to search out legitimate avenues to address the problem. Not surprisingly, members of public interest organizations are, in general, highly efficacious individuals; they vote regularly, contribute to political parties and candidates, write letters to public officials, and are politically informed. With these characteristics and with disproportionately higher levels of education and income, these individuals' membership in public interest organizations may be viewed as simply another manifestation of their political efficacy and activism. Therefore, in the absence of additional motivations, one would expect to find a strong consistent motivational pattern based on goals and threats among members of public interest groups.[18]

The distinct organizational goals and incentives offered by each group competing for membership aid potential members in discerning the most optimal organization or organizations to support. Contrary to Smith's theoretical analysis of environmental group members, individuals *do not* make mutually exclusive choices when becoming active in the environmental movement.[19] In fact, membership in several groups is the norm rather than the exception among environmentalists. Given the limited information provided to potential members through direct marketing efforts, these individuals must rely on perceptions of the groups and their goals that may lead to a pattern of overestimation of the groups' collective goals in much the same manner as one might overestimate one's efficacy. In spite of these barriers, there still may remain a distinct set of motivations for belonging to a particular group, based on the perceived or actual goals of the group.

Apart from the research of Moe, Rothenberg, and Hansen, there is little empirical research that directly addresses the impacts of organizational incentive structures on the motivations of individuals who belong to public interest organizations. Consequently, the validity of the analysis to be presented in this chapter depends on the degree to which each of the dimensions presented above is defined and operationalized in a substantively and theoretically accurate manner. Culling from more than 350 questions asked of each survey respondent, several questions were used to construct each of the four motivational dimensions.[20]

Organizational Incentives and Individual Motivations

In order to maintain themselves and continue to be competitive in the environmental marketplace, environmental organizations must offer distinctive sets of incentives to attract new members and retain existing ones. Individuals sufficiently attracted by some combination of incentives offered by a particular group, assuming willingness and ability, will then support it.

Fortunately, the membership surveys included questions directed specifically at the group from which respondents were sampled, regarding individuals' motivations for belonging to each particular environmental group. (For example, Environmental Defense Fund members were asked why they belong to EDF, not why they belong more generally to environmental groups.) Hence, their responses should not be generalized into motivations attributable to environmental groups in general.

The findings presented in figure 5.1 offer evidence of the diversity of incentives and motivations in environmental organizations. The dashed lines

FIGURE 5.1 Membership Motivations for Belonging to Environmental Groups (Percentage of Group Members Considering Each Incentive to be Important in Their Decisions to Belong)

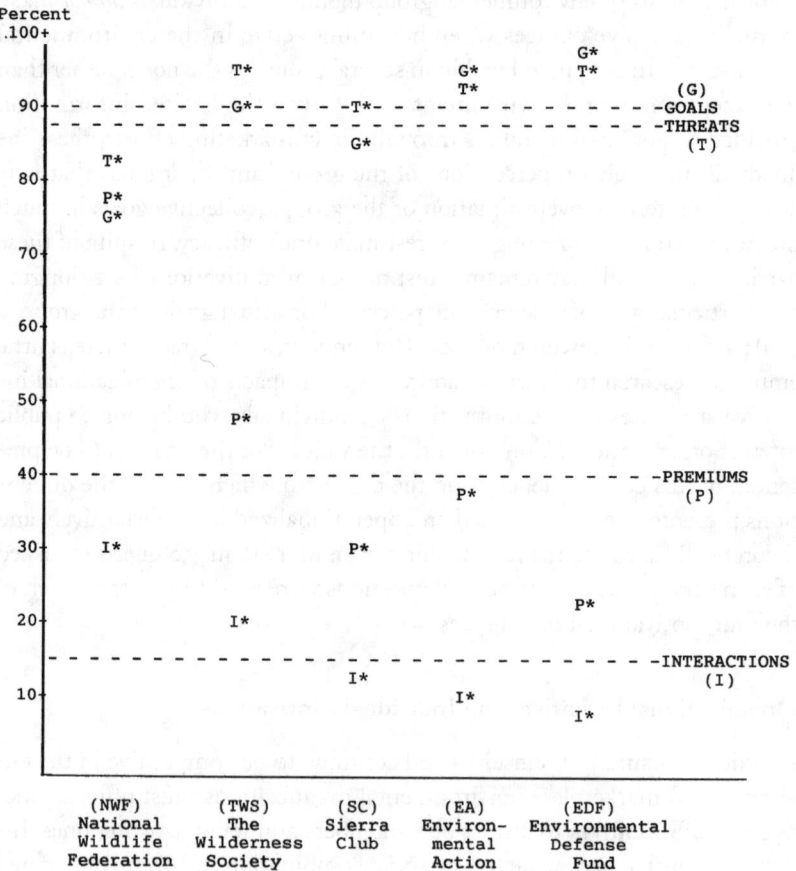

represent the aggregation of members' motivations, while the group data points represent the percentage of members who find that each incentive dimension is important to their decision to belong to each organization. Obviously one arrives at a rather different assessment of group incentives and individual motivations when individual motivation patterns are assumed not to be group-specific. It is evident that these five environmental organizations do provide distinct combinations of incentives that attract a certain portion of environmentally attentive citizens by appealing to some combination of individual motivations.

While it is clear that individuals within the national environmental movement exhibit patterns of belonging to more than one organization at a time, each individual's membership in a particular environmental group is not a chance proposition. By the same token, the dynamic of membership in an environmental organization does not end at the group level. Individuals may belong to the same group at the same time for different reasons. This pattern, of course, depends upon the mix of incentives offered by a particular group. Realistically it would be difficult for an organization that offers only its substantive goals as incentives, with few premiums and little opportunity for group interactions, to attract a motivationally diverse membership. Yet it may be expected that the greater the mix of incentives provided, the more motivationally diverse the organization's membership.

Though each group provides a distinct mix of incentives, the dimensions of the incentives are similar across groups. For as demonstrated in table 5.1, members of environmental groups are fairly consistent in their receptivity to goals and threats. However, these dimensions do not incorporate any specific policy direction. National environmental organizations encompass a broad range of concerns from the more conservative-oriented issues of wilderness protection to the more environmental issues of nuclear power and toxic waste management. As expected then, these organizations attract a distinctive, yet varied membership base (see table 5.1).

Before analyzing the prospect that the incentive mix affects the motivational diversity of the membership, some explanation is necessary regarding the incentives offered by each of the five organizations in 1978, at the time the survey was conducted. The National Wildlife Federation is comprised of a national organization and fifty-one state federations. In many cases, the clubs and nature conservation groups provide members with the opportunity to meet and socialize. NWF also offers a wide range of publications (including *National Wildlife, International Wildlife, Ranger Rick,* and *Your Big Backyard*), clothing and book discounts, and vacation and travel opportunities. In addition, it operates with an extensive conservation

TABLE 5.1 Characteristics of Environmental/Conservation Group Members

	ENVIRONMENTAL ACTION (N = 705)	ENVIRONMENTAL DEFENSE FUND (N = 630)	NATIONAL WILDLIFE FEDERATION (N = 509)	SIERRA CLUB (N = 661)	THE WILDERNESS SOCIETY (N = 623)
Education					
High School Graduate or Less	7%	6%	34%	7%	9%
Some College	13%	11%	23%	16%	14%
College Graduate	80%	83%	43%	77%	77%
Income					
Under $10,000	15%	7%	17%	9%	13%
$10,000–$19,999	29%	24%	38%	31%	26%
$20,000–$39,999	33%	38%	36%	41%	36%
$40,000 and Over	23%	31%	9%	19%	25%
Median Age	39	46	51	38	47
Sex					
Male	63%	58%	54%	68%	57%
Female	37%	42%	46%	32%	43%

Party Identification					
Democrat	31%	29%	12%	29%	20%
Independent	65%	62%	64%	59%	64%
Republican	4%	9%	28%	12%	16%
Political Ideology					
Conservative	9%	13%	46%	23%	29%
Middle-of-the-Road	9%	13%	31%	16%	18%
Liberal	72%	67%	21%	57%	51%
Radical	10%	7%	2%	4%	2%
Religious Affiliation					
No Organized Group/None	56%	47%	22%	54%	40%
Protestant	27%	34%	60%	29%	45%
Catholic	6%	8%	16%	11%	10%
Jewish	11%	11%	2%	6%	5%

policy agenda, although it is not clear how members could know the extent of the policy effort, given the limited policy content in the communications sent to them. The content of the publications of each of the five organizations will be discussed in detail in the next chapter. The Sierra Club also provides all three group-controlled incentives:[21] interactions—organization contacts at the national, regional or state (chapter), and local (group) levels, numerous meeting opportunities, outings program, and travel packages; premiums—an attractive and somewhat informative national magazine, newsletters from chapter and group organizations, discounts on books, clothing, and equipment, and recently a Sierra Club Visa card; and goals—extensive environmental and conservation agenda formulated through policy inputs from all organizational levels. The Wilderness Society is organized at the national level only. The organization does provide premiums in the form of a magazine (and a newsletter until 1982). At the time of the survey, an outings program was offered, but it generated little response and has since been eliminated. The policy agenda was clear and focused almost exclusively on wilderness issues and was articulated to its members.

Environmental Action was also organized in the nation's capital only. EA offered a fairly brief, twenty-four-issue magazine (later cut back to quarterly and expanded) as a premium. It offered few opportunities for interactions and the goals, in the form of a policy agenda, appealed only to the more liberal (and some would argue radical) wing of the environmental community. The Environmental Defense Fund offered the closest to a purely goal-driven incentive structure at the national level. The organization's goals were its marketing tools at the time of the membership survey. Since this organization has a distinctive approach to environmental policy making—scientific and economic research and litigation—it has attracted a more specialized, attentive membership. As for premiums, it offered very little—a quarterly, four-page, two-color newsletter (most recently expanded to eight pages, bimonthly). Indeed, EDF leaders marketed their group as an anomaly. Recall that in a membership renewal letter, Executive Director Frederic Krupp concluded: "No canvas bag, no bumper sticker for continuing with us. Just the knowledge that you will help reduce our country's acid rain by more than 50 percent."[22]

With these incentive patterns in mind, the results of a discriminant analysis are presented in table 5.2 that reveal the degree to which each organization attracts a distinct membership based on its incentive structure. If you have ever filed federal income taxes, you have been subjected to a discriminant analysis. The Internal Revenue Service conducts its audits

TABLE 5.2 Discriminant Analysis (Individual Members Classified by Four
Motivational Dimensions and Policy Predispositions)

PERCENTAGE OF MEMBERS CLASSIFIED
INTO GROUPS:

Actual Group:	EA	EDF	NWF	SC	TWS
Environmental Action	*26.56*	28.52	16.04	11.76	17.11
Environmental Defense Fund	17.38	*52.56*	5.52	9.82	14.72
National Wildlife Federation	6.76	3.24	*68.92*	6.76	14.32
Sierra Club	18.00	24.36	16.55	*18.91*	22.18
The Wilderness Society	15.74	21.70	22.77	11.49	*28.30*

based on this methodology. The IRS is able to search its entire tax return database on a variety of criteria in order to select out likely audit candidates. If, for example, you claim eight or more dependents or your deductions exceed 44 percent of your income, the discriminant analysis conducted by IRS computers will select your return as a possible candidate for an audit. The same procedure may be applied to all 3,128 survey respondents.

Instructing the computer to place similar respondents in the same organizational category based on the four motivational dimensions—premiums, goals, interactions, and threats—as well as two self-descriptive measures of individuals' general policy predispositions,[23] leads to results offering some interesting insights into the interrelationships between organizational incentives and membership motivations across these five environmental organizations.[24]

The results of the analysis offer mixed evidence of the expectation that a greater mix of incentives provided by the group results in a more motivationally diverse membership. Perhaps the easiest way to interpret the results in table 5.2 is to look first at the italicized percentages on the diagonal from top left to bottom right. If each group offered a distinct mix of incentives that equally motivated its members, then the diagonal percentages would register 100 percent and the remaining cells would be empty (0 percent). Obviously this is not the case. Nevertheless, in two instances (EDF and NWF) more than half of the observations are correctly paired with the groups from which they were sampled. The motivation patterns of EDF members are more easily interpreted because EDF respondents are not offered a motivationally diverse set of incentives. Just over half of EDF members (52.6 percent) are correctly arrayed on the basis of individuals' motivations and predispositions. An additional 17.4 percent fall into the EA grouping. EA shares with EDF the policy concerns of the environmental agenda and, therefore, attracts like-minded members on this dimension. In fact, 40 percent of EA members in the sample belong to EDF as well.

Almost 70 percent of NWF members are correctly classified. Concerning an organization with multiple incentives, one would expect that organization members would be less concentrated on motivational dimensions vis-à-vis members of other organizations. However, these findings suggest that while multiple incentives are offered, members are motivated to belong to the group only by one of the incentives. As the analysis presented in figure 5.1 shows, NWF members are distinctive in their strong attachment to premiums and, to a lesser degree, group interactions. At the same time, however, the motivational patterns of the Sierra Club and The Wilderness Society members do suggest the diversity of individual incentives and motivations, particularly in the case of the Sierra Club. Only 18.9 percent of SC members are correctly classified; further, they are distributed fairly evenly across the other groups. The evidence is supported by the fact that the Sierra Club membership list is one of the most sought after among national environmental organizations, because so many groups have found that the SC lists work very well for them in recruiting new members.[25]

Similarly, TWS members fall into NWF ranks perhaps on policy direction and into EA based on similarities on the premiums dimension. One would assume that in reality there would be less overlap here due to the very different policy agendas of each organization. Finally, EA members fall almost equally into their correct category and into EDF. Again, the policy congruence between these groups is evident. Nevertheless, in light of these diverse findings, one might now be a bit more cautious in attributing motivations to individuals based on the mix of incentives offered by organizations.

Of the groups sampled, the National Wildlife Federation numbered at the time of the survey more than 620,000 associate members and more than 3,500,000 affiliated members. The remaining group memberships in the sample range from 178,000 (Sierra Club) to 16,000 (Environmental Action). It is possible to analyze the relative importance of selective incentives or premiums in maintaining large public interest organizations vis-à-vis smaller groups and provide some empirical insight into Olson's argument regarding size and selective incentives. Consequently, based on Olson's argument, it may be anticipated that the provision of nongoal-oriented incentives is necessary for the maintenance of large environmental organizations.

TABLE 5.3 Logistic Regression Analysis Membership Motivations: Members of Large Groups (NWF) Compared with Members of Smaller Groups

DEPENDENT VARIABLE: 1 = NWF MEMBERS; 0 = OTHER GROUPS MEMBERS

Variables	Beta (MLE)	Std. Error	Chi-Square	P
INTERCEPT	−4.104	0.528	60.48	.0000
PREMIUMS***	1.640	0.104	248.93	.0000
GOALS*	−0.592	0.115	26.62	.0000
INTERACTIONS**	0.408	0.063	41.90	.0000
THREATS	−0.478	0.117	16.64	.0000

Model Chi Square = 565.17 (p < 0.001) Model R^2 = .252
Correct Predictions = 83.9% (N = 2,563)

The results presented in table 5.3 provide some indication of the validity of Olson's size argument by employing a logistic regression analysis due to the binary categorical nature of the dependent variable in each case.[26] The dependent variable in this analysis separates members of the largest group in the environmental/conservation movement, the National Wildlife Federation, (coded 1) from members of the remaining four groups (coded 0). Independent variables include the four motivational variables—premiums, goals, interactions, and threats.

In this regression analysis, the significant results (p <= 05) in tables 5.3–5.6 are presented in italics. In addition, three asterisks (***) are placed next to the most significant variable in each model, followed by two asterisks after the next most significant variable, and one asterisk after the third most significant. For these tables, the correct predictions result may be interpreted as the degree to which the variables in each model accurately predicted whether each survey respondent belonged in dependent variable category 1 or 0. Finally the R-squared result presents the percentage of the variance in the overall relationship between respondents in categories 1 and 0 explained by the variables in each model.

Clearly, the motivational patterns of members of the largest environmental groups are distinguishable from those of members of smaller organizations, primarily on the premiums dimension and secondarily on the interactions dimension, but also more weakly (and negatively) on the two remaining dimensions. These findings thus support Olson's argument and are consistent with the nonpolicy mobilization efforts of NWF reported in chapter 3. Further, members are mobilized through social network means, that is through friends, personal contacts, or other interpersonal contacts.

Social network recruits place greater emphasis on premiums and interactions than do direct-mail recruits.[27] The significant negative relationships identified for both goals and threats motivations point to the comparatively lower levels or group-specific purposiveness and a diminished sense of movementwide efficacy among NWF members.

The National Wildlife Federation membership is distinct from the other environmental memberships in its almost total reliance on premiums as incentives. While the organization may maintain its membership by appealing (albeit rather vaguely) to policy goals as motivating factors, it makes little effort to place the goals of the group before the membership. NWF leaders have made a conscious decision not to broadcast to their members the details of the organization's policy concerns, mainly for fear of losing portions of its vast membership base. Consider, for example, the nuclear power issue. In the case of Environmental Action, one may easily predict the sentiment among its members on this issue (strongly opposed), mainly because the leaders, in their communications, wear their collective hearts on their sleeves. If one wished to learn about the policy direction of EA, all one needed to do was read its magazine. This is not the case for NWF. According to the issue responses of NWF members in the sample, the membership is split on the issue of nuclear power. The leadership had taken a position on the issue (antinuclear); they simply did not focus on the issue, fearing the loss of the supporters of nuclear power.

One additional illustration highlights the consequences of an incentive system based on premiums and interactions. If one remembers the tenure of James Watt as secretary of the interior, one will also remember the utter outrage among environmental activists and their campaigns for Watt's removal. One of the last environmental and conservation organizations to speak out against Watt was NWF. The decision to do so was not made hastily. In fact, NWF commissioned an extensive mail survey of its associate members (800,000 in 1981) and affiliate leaders (Sample N = 2,626). Only when the survey results showed members' issue opinions in opposition to the views of the secretary of the interior did the leadership make a public statement in support of Watt's removal.[28]

The results of this membership survey also suggest the possibility that environmental groups stressing non-goal-oriented incentives attract a more mass-oriented membership than do groups placing little or no emphasis on premiums and interactions. Salisbury has argued that groups stressing these material incentives over policy goals are most clearly operating within an "exchange" framework where selective incentives are provided by leaders in exchange for individual contributions.[29] Thus when the premium is attractive and the relative cost is reasonable, it is not necessary to appeal to goals and threats as motivations in order to attract a substantial number of members. Implicitly, direct economic benefits appeal to the most basic level of consciousness and, hence, are attractive to a broader range of potential members. At this most basic level, it may be more accurate to characterize such individuals as subscribers rather than members.[30]

In the 1981 NWF survey respondents were asked a series of questions concerning their personal views on a range of conservation issues. On these questions, most respondents disagreed with the positions of Watt, although almost half of the respondents were unaware of their disagreement. When given a series of statements about Watt's policies toward conservation and development, 44 percent of the associate members chose "I don't know enough about his policies to make a judgment." While no comparative data are available for members of other groups, it would seem that the salience of the Watt issue in June of 1981 and the attentiveness of the environmental constituency should have generated higher levels of knowledge among environmental organization members than was manifest among NWF members.

The results presented in table 5.4 point to the significant differences in demographic characteristics and personal attributes between members of

TABLE 5.4 Logistic Regression Analysis Membership Demographic
Characteristics and Attributes Members of Large Groups (NWF)
Compared with Members of Smaller Groups

DEPENDENT VARIABLE: I = NWF MEMBERS; O = MEMBERS/OTHER GROUPS

Variables	Beta (MLE)	Std. Error	Chi-Square	P
INTERCEPT	6.329	0.934	44.65	.0000
EDUCATION***	−0.558	0.108	26.30	.0000
INCOME	−0.169	0.081	4.38	.0363
PARTY	−0.283	0.072	15.21	.0001
IDEOLOGY	−0.287	0.104	7.57	.0059
RELIGION*	0.890	0.213	17.71	.0000
ENVIRON	−0.435	0.190	5.21	.0225
CONSERV	0.224	0.198	1.27	.2591
MAIL	−0.758	0.222	11.72	.0006
EFFICACY	−0.407	0.184	4.90	.0269
OCCUP	−0.313	0.280	1.24	.2650
NUMGRPS**	−0.355	0.074	22.79	.0000

Model Chi-Square = 329.89 (p < 0.001) Model R^2 = .318
Correct Predictions = 87.5% N = 1,028

large environmental groups (NWF) stressing premiums and interactions and members of smaller, more goal-directed groups.[31] In general, members motivated by nonsubstantive incentives have a greater probability of being less educated, less wealthy, more conservative and Republican, members of an organized Judeo-Christian religion, and also have a greater probability of *not* being "environmentalists," direct-mail recruits, and members of several groups simultaneously.[32] Finally, these members are less likely to be as politically efficacious as members of other groups, having more in common with the general public than with their more elite environmentalist counterparts.

The results are even more striking when members belonging only to NWF (48 percent of NWF respondents belong to no other environmental or conservation group) are analyzed. These single-group NWF members are almost totally motivated by premiums. The remaining 52 percent (multiple members) do identify with goals and threats as motivating factors, but not to the extent of members in the control group. It is quite realistic to assume that individuals motivated by goals and threats who belong to one or more goal-directed organization may also belong to NWF (or perhaps to the National Audubon Society) simply to receive an entertaining, colorful magazine for themselves or *Ranger Rick* and *Your Big Backyard* for their children—premiums not available from the more goal-directed groups.

In addition to the multiple incentives offered by environmental groups that may attract potentially distinct members, membership-oriented organizational structures and activities that reach beyond the group's national headquarters also may provide inducements for members to maintain their group affiliations. State, local, and regional suborganizations provide an additional connection between leadership and membership and may promote a greater degree of membership interaction and input. Group interactions at the subnational level such as outings, vacation packages, hiking, biking, and camping also offer affective inducements to group members. Thus, members of organizations with membership-oriented organizational structures and activities beyond the national level may also have higher levels of affective commitment than members of groups without such structures and activities.

Three of the five groups offered structures (actual subnational organization: meetings, elections, leaders) and/or organizational activities at the time of the survey that may have induced a greater interactive commitment to the group: NWF, SC, and TWS. First, in model 1 of table 5.5, NWF and SC are paired, as these groups are organized for membership purposes beyond the national level. Model 2 includes TWS because this group did

TABLE 5.5 Logistic Regression Analysis Membership Motivations:
         Organizational Structures and Activities

Model 1: Affective Differences Between Members of Groups with Organizational Structures Beyond the National Level and Members of Nationally Organized Groups

DEPENDENT VARIABLE:1 = NWF/SC MEMBERS; 0 = OTHER GROUP MEMBERS

Variable	Beta (MLE)	Std. Error	Chi-Square	P
INTERCEPT	1.074	0.367	8.54	.0035
PREMIUMS***	0.424	0.056	57.72	.0000
GOALS**	−0.579	0.082	49.73	.0000
INTERACTIONS*	0.194	0.045	18.26	.0000
THREATS	−0.223	0.085	6.87	.0087

Model Chi-Square = 242.10 (p < 0.001) Model R^2 = .069
Correct Predictions = 66.8% (N = 2,563)

Model 2: Affective Differences Between Members of Groups with Organizational Structures and/or Activities Beyond the National Level and Members of Groups Organized at the National Level

DEPENDENT VARIABLE: 1 = NWF/SC/TWS MEMBERS;
         0 = OTHER GROUPS MEMBERS

Variable	Beta (MLE)	Std. Error	Chi-Square	P
INTERCEPT	1.004	0.376	7.14	.0075
PREMIUMS***	0.511	0.054	88.64	.0000
GOALS**	−0.626	0.085	54.43	.0000
INTERACTIONS*	0.182	0.045	15.95	.0001
THREATS	−0.017	0.086	0.04	.8481

Model Chi-Square = 255.21 (p < 0.001) Model R^2 = .071
Correct Predictions = 67.0% (N = 2,563)

provide an outings and activity program for its members. On the interactions dimension, the maximum likelihood estimate for the first pairing (NWF and SC) is slightly more significant than with the inclusion of TWS; yet both models provide evidence that those groups with subnational organizational structures and activities have members with higher levels of commitment on this dimension. However, the premiums (+) and goals (−) dimensions remain stronger predictors. It is also important to note that on the threats dimension, much of the negative relationship is attributable to NWF only, as a comparison with the estimate in table 5.3 demonstrates.

Just as those groups disproportionately relying on premiums attract a more broad-based membership, one might expect to find members of groups offering only goals as incentives to attract a much more elite membership. In addition, one would expect that these members are also affiliated with other related organizations. It may be the case that once an individual belongs to one or more organizations that supply interesting and attractive magazines and other premiums, he or she may then seek out more specifically goal-driven organizations without the need for attractive premiums. Membership in just four or five groups may easily cost $100-$200 per year, if one gives only slightly more than the minimum request for dues. Obviously, it is necessary to have disposable income in order to "participate" more widely in the national environmental movement. In addition, length of involvement and receptivity to direct mail go hand in hand. The longer one belongs to an organization or organizations, the greater the chance of one's name circulating on mailing lists to other similar (and not so similar) groups. Virtually all groups are using some form of direct-mail proselytizing. If an individual simply discards direct-mail solicitations, the possibility of joining additional groups (e.g., through social network means) is greatly diminished.

Of the groups analyzed only one concentrated disproportionately on a single goal-oriented incentive strategy to attract and maintain its members, the Environmental Defense Fund. Table 5.6 provides comparisons of this single incentive group to other groups and evidence of the elite and multiple membership nature of such organizations. Model 1 indicates that the probability of belonging to EDF is enhanced by members' lack of motivation for premiums and to a lesser extent by their stronger goal-oriented motivations. In terms of the elite characteristics of membership, Model 2 offers some support for demographic differences but identifies two additional variables that are stronger probabilistic indicators of membership in such groups.

EDF members are more likely to be better educated, Democratic, less predisposed toward conservationism, more politically efficacious, and more likely to work in occupations of higher status. However, multiple-group membership and direct-mail recruitment are far better predictors of membership in groups with only a single incentive.[33] In the sample, only 4 percent of EDF members belong to no other groups. These findings intuitively make sense. Organizations marketed solely through their policy goals, by their very nature, place little emphasis on selective economic benefits and consider the effort of maintaining a membership to be a "necessary evil." Consequently, they rely on the easiest and most cost effective means—di-

TABLE 5.6 Membership Motivations and Attributes: Pure Incentive Group Members and Mixed Incentive Group Members

Model 1: Membership Motivations

DEPENDENT VARIABLE: 1 = EDF MEMBERS; 0 = OTHER GROUP MEMBERS

Variables	Beta (MLE)	Std. Error	Chi-Square	P
INTERCEPT	−2.288	0.510	20.06	.0000
PREMIUMS***	−0.925	0.071	170.71	.0000
GOALS**	0.566	0.113	25.06	.0000
INTERACTIONS*	0.122	0.059	4.21	.0402
THREATS	0.230	0.116	3.89	.0485

Model Chi-Square = 281.28 (p < 0.001) Model R^2 = .107
Correct Predictions = 72.1% (N = 2,563)

Model 2: Membership Characteristics and Attributes

DEPENDENT VARIABLE: 1 = EDF MEMBERS; 0 = OTHER GROUP MEMBERS

Variables	Beta (MLE)	Std. Error	Chi-Square	P
INTERCEPT	−6.406	0.890	51.75	.0000
EDUCATION*	0.294	0.112	6.86	.0088
INCOME	0.043	0.057	0.57	.4486
PARTY	0.156	0.065	5.71	.0169
IDEOLOGY	0.145	0.093	2.42	.1201
RELIGION	−0.099	0.167	0.35	.5561
ENVIRON	0.100	0.181	0.31	.5796
CONSERV	−0.365	0.181	4.08	.0434
MAIL**	1.530	0.186	68.03	.0000
EFFICACY	0.265	0.153	2.99	.0838
OCCUP	0.817	0.338	5.85	.0156
NUMGRPS***	0.322	0.032	101.66	.0000

Model Chi-Square = 389.85 (p < 0.001) Model R^2 = .274
Correct Predictions = 85.0% (N = 1,187)

rect mail. What distinguishes these groups from mixed incentive groups is their application of direct mail without the significant aid of premiums. Given the strategy of direct mail, the individuals being recruited by an environmental organization are very likely to be members of at least one public interest organization and, more specifically, judging from the patterns of list rentals (see table 4.3), likely to be members of a national environmental organization.[34] Direct-mail recruits, then, are socialized, to an extent, into the direct-mail process and into the public interest network. An individual particularly concerned with environmental issues, such as acid rain and toxic waste management, interested in the more scientific aspects of environmental policy making or the legal avenues of policy influence, and already a member of one or more related organizations (perhaps providing premiums), may then decide to join an organization with a single (goals) incentive. An unfortunate consequence of providing only policy goals as incentives is the difficulty of the communications aspect of the organization; EDF had the lowest retention rate of first-year members among the five groups—35 percent. This suggests that some offering of premiums facilitates the maintenance of environmental organizations and public interest organizations more generally.

The preceding discussion of the direct-mail recruitment phenomenon relates directly to the pattern of multiple membership identified among environmental respondents. Table 5.7 provides a demographic/attitudinal/behavioral model of multiple membership. The analysis presented em-

TABLE 5.7 Multiple Regression Analysis Membership Characteristics:
Multiple Membership

DEPENDENT VARIABLE: NUMBER OF GROUPS

Variables	Parameter Est.	Std. Error	T-Score	P
INTERCEPT	−8.262	0.886	−9.32	.0001
EDUCATION	0.038	0.095	0.39	.6935
INCOME**	0.355	0.056	6.38	.0001
PARTY	0.112	0.059	1.89	.0589
IDEOLOGY	0.229	0.085	2.71	.0068
RELIGION	−0.064	0.169	0.35	.7568
ENVIRON	0.509	0.167	3.05	.0023
CONSERV	0.457	0.166	2.76	.0059
MAIL***	1.850	0.158	11.69	.0001
EFFICACY	−0.058	0.147	−0.40	.6911
OCCUP	0.411	0.267	1.54	.1247
ACTIVITY*	0.597	0.117	5.07	.0001
SERIOUS	0.341	0.114	3.00	.0028
SEX	−0.115	0.169	−0.68	.4969
AGE	0.041	0.006	1.54	.1247

F VALUE = 39.410 ($p < 0.001$) R^2 = .273 ADJ R^2 = .266

ploys a standard multivariate linear regression model; the dependent variable is continuous—number of groups to which the respondent belongs.[35] These results indicate that income, direct-mail membership, and political activism are the most significant variables in the model. Other significant explanatory variables include members' perceptions of the seriousness of environmental problems, predispositions toward both environmentalism and conservationism, and liberal ideology. Interestingly, the level of education and general political efficacy are particularly weak indicators of membership in multiple groups. The latter is a function of opportunity (direct mail) and ability (income), but also a substantive concern.

Up to now, this chapter has explored the theoretical questions directed at the interactive relationships between organization incentives and individual motivations within various environmental organizations. One final internal relationship, however, remains to be investigated. Prior to the work of Rothenberg, existing models and theories assumed no variation in individual motivations within an organization over time—that an individual retains the same motivational patterns from the minute he or she decides to join until the time of departure from the group. Given Rothenberg's findings in his analysis of Common Cause members, it is plausible to assume that an individual may be motivated by a certain mix of incentives to join a group. Assuming an acceptable level of satisfaction, the individual renews his or membership regularly. As time passes, his or her cumulative exposure to the organization and its goals causes the person to reevaluate gradually his motivations for belonging.

Since there are only organization members in the five samples, one may assume that at the time of their last renewal (or at the point that first-year members joined) the members were sufficiently satisfied with some relevant aspect of the group. While premiums and interactions may play an important role in attracting new members and even maintaining them after the crucial first year (given the limited information available at the outset concerning the group's means and ends), it is plausible that agreement with the organization's policy may become the maintaining force as length of membership increases. The sample respondents under consideration have expressed motivations "for belonging to" a particular environmental group; perhaps those members with relatively short tenures in their organizations (one year or less) would offer little discernible difference in motivations for "joining" and motivations for "belonging." However, one should not assume a static relationship between individual motivations for joining a group and more long-term motivations for belonging to it.

As the preceding analyses demonstrate, membership motivations are not easily generalized. Individuals belong to different (yet similar) groups for a variety of reasons. Further, individuals may be motivated to belong to the same group by different incentives. The final analysis of the nature of environmental organization membership incorporates time into the internal organizational dynamic. Optimally, the data needed to test changes in individual motivation patterns would be gathered through long-term panel surveying—interviewing the same individuals several times over a period of several years. Unfortunately, only cross-sectional data are presently available, thus future research is suggested on this particular question.[36]

The preceding analyses have uncovered several new relationships between individuals and environmental interest organizations. Public interest organizations have distinctive incentive systems and, in most cases, incorporate a mix of incentives. There is evidence to suggest, however, that while groups offer a range of incentives, they may (as is the case with Sierra Club) or may not (as with National Wildlife Federation) attract a motivationally diverse membership. In terms of group size, there is tentative support for the necessity of premiums in maintaining large public interest organizations. In addition, the provision of premiums has an impact on the demographic and attributional composition of the membership. Membership-related structures and interactions also generate greater affective commitment among members. Conversely, members of groups with single incentives are found to be demographically more elite than members of groups with mixed incentives and are also much more likely to belong to several groups. The number of groups to which one belongs is related to the individual's opportunity to join other groups, ability to contribute, and substantive concerns.[37]

In summary, this attempt at identifying the complex relationships between the incentives of environmental organizations and the motivations of their individual members yields some interesting findings. The motivational dimensions are provisionally valid and reliable, and the results are quite useful, at least in identifying and testing some of the relevant components of the relationships and in laying to rest some of the assumptions that impede our understanding of the nature of membership in public interest organizations. Given these findings, it is virtually a necessity that organizational leaders make a more concerted effort to learn about their members—why they join, why they remain loyal to the organization, and perhaps most important, what they think about the policies pursued by the organization.

6

The Heart of the Matter
Leadership–Membership Connections

> If they [organization leaders] are sending out good communications, to some degree, they are going to wind up talking to themselves. There is a real danger, if you have an undifferentiated mass of members out there, where you have a one-on-one relationship—you to them. Who are you talking to? Or are you just hearing the echo of your own voice?
>
> —JOEL THOMAS, GENERAL COUNSEL, NATIONAL WILDLIFE FEDERATION

As we have discovered in the preceding chapter, the concept of membership in national environmental organizations has multiple meanings. While in some instances people who join environmental as well as consumer, senior citizens, single-issue, or other types of public interest organizations may attribute the same level of significance to this type of membership as to their memberships in churches, social and/or fraternal organizations, or in occupational groups, most do not. A growing number of citizens have little more than what Michael Hayes has identified as "checkbook affiliations" with public interest organizations.[1] With this in mind, however, we should not discount totally the link between the leaders of public interest organizations and their followers or adherents, for, as John Gardner argues, this connection is "the heart of the matter" for organizational leadership.[2]

One aspect of membership that concerns leaders of organized interest groups is the ability of their members to act as advocates in the larger political arena, to echo their leadership's voices. This concern is not at all new to interest group leaders. Schlozman and Tierney have called grassroots lobbying, the mobilization of constituents for political purposes, "an ancient weapon in the pressure group arsenal."[3] Indeed, since the founding of the republic, economic interests and later social interests have echoed the demands of their leaders to officials in Washington through adherents in the constituency.[4] In the modern era of mass communications and direct mail, lobbying organizations have undertaken systematic mobilizations of their members to create public pressure for their legislative agendas.[5] Letter-writing campaigns, telephone alerts, and other grassroots techniques are now standard items in the interest group arsenal, and most groups employ these tactics with greater frequency than in the past.[6]

Organizational leaders rely on members back home to press their case because such expressions of support legitimize their claims. Lawmakers who stoutly resist the arguments of "special interests" will give the same views serious attention if they have a constituency connection. As elected representatives, they have a special obligation to heed citizens who take the trouble to voice an opinion, and most members of Congress tend to place some significant degree of emphasis on constituent sentiments in making legislative decisions. In a 1986 study of 147 House offices regarding the processing of mail from constituents "almost two-thirds (62.2 percent) of the [House] members are reported to ask about legislative mail from constituents prior to voting on substantive policy issues on a regular basis; an additional 30 percent ask about the mail 'occasionally.'"[7] A more recent study of 150 House and Senate offices, conducted in 1993, recorded similar responses as more than 70 percent of the members paid close attention to "non-form, personally written letters from constituents in the home district or state."[8] Efforts to mobilize the membership, then, are generally perceived on Capitol Hill as being rather effective.[9] In the words of one lobbyist, "We really put a lot more stock in what people can do at home to influence members of Congress than in what we can do by talking to congressmen."[10]

Despite the importance given to the mobilization of members for political influence by interest group leaders as well as legislators, there is very little understanding of how or why it works—if it does, in fact, work. Several recent studies of interest groups indicate that organization leaders often are uncertain about the effectiveness of their efforts to rouse their members. Some concede that they use direct-mail alerts because they have invested in sophisticated computer technology and consider it wasteful not to use it.[11] Others express doubts about the reliability and consistency of their members' support.[12] An official of the U.S. Chamber of Commerce—a group noted for its strong emphasis on personal contacts between members and lawmakers—summed up the many uncertainties of membership mobilization activity when he noted that assessing its impact was like "a blind man searching for a black cat in a coal bin at midnight."[13]

This chapter seeks to address some of the ambiguity surrounding the strategies of the membership's political participation in the context of the national environmental movement. At its best, organized membership and leadership participation in the policy-making process incorporates (and integrates) members' preferences and varying intensities of preferences with organizational interests. Among the memberships of the five environmental organizations presented in the preceding chapters, less than one-fifth of the members participate in issue alerts or have their names on special mailing or telephone lists, yet almost two-thirds of the members write letters to members of Congress and administration officials on environmental matters without direct stimulation from organization leaders (see table 6.1).

In order to focus on the dual nature of membership political participation—as *interest group advocates* and *citizen activists*[14]—the more indirect means through which leaders of environmental organization are able to cue their members to the more salient issues on their particular environmental agendas, namely, the communications (magazines, newsletters) sent to group members by the organization leaders are analyzed. Communications sent to all members of the five organizations under study for two periods in the 1970s (1977–1978) and the 1990s (1991–1992) were analyzed for their content. Following the content analysis of the organizations' communications (approximately 3,100 pages), the degree of cohesiveness and breadth of the members' issue agendas were examined for the earlier period with implications outlined for contemporary interest group interactions.

Environmental organizations provide an excellent opportunity for evaluating the nature of the relationships between organization leaders and active citizens who belong to environmental groups. Throughout the bitter struggles over landmark legislation passed during the 1970s, observers of the congressional scene credited environmentalists' mobilizations at the constituency level with the unprecedented success in defeating formidable coalitions of business and labor interests arrayed against them.[15] This interpretation of the legislative record is credible for two reasons. First, individuals who joined environmental groups tended to be affluent, well-educated, and politically active. One obvious manifestation of the level of political activism is belonging to public interest organizations such as environmental groups.[16] These are the types of citizens who typically dominate congressional politics at the constituency level. Second, financial constraints have prevented these groups from establishing extensive Washington operations (although there has been significant change in recent years).[17] Even today, because of their comparative financial disadvantages, public interest organizations rely heavily on alternative tactics of influence,

TABLE 6.1 Membership Roles and Activities

	Environmental Action	Environmental Defense Fund	National Wildlife Federation	Sierra Club	The Wilderness Society
% Members who consider themselves to be "Active" in the Interest Group	6%	2%	4%	10%	5%
% Members who don't think of themselves as Members of the Interest Group	44%	58%	35%	25%	39%
% Members who had formed their opinions on environmental issues prior to joining the Interest Group	94%	94%	80%	88%	88%
% Members who have read (envir. magazines) at least occasionally during the past year	99% (83%)*	98% (69%)	95% (65%)	99% (75%)	100% (83%)
% Members who are on special mailing lists and respond to issue alerts by writing to their members of Congress	19%	11%	8%	23%	26%
% Members who have contacted a public official about an envir. matter during the last year	64%	74%	46%	56%	67%

*Percentage of members reporting "frequent" readership.

such as coalition formation, staged media events, and constituency mobilizations. This multifaceted strategy continues into the late 1990s. Judging from the survey of interest groups by Schlozman and Tierney, membership mobilization remains the centerpiece of the policy influence game.[18] Thus, the events of the past twenty years, the attributes of the memberships, and the strategic calculations of the organization leaders suggest that environmental organizations provide a fitting context for the inquiry into the political participation of organized constituents.

Limits and Possibilities of Organized Constituents

For more than thirty years, Bauer, Pool, and Dexter's conclusions about the limits of constituency pressure have served as the definitive statement on the subject.[19] Their research on tariff policy demonstrated that a variety of barriers prevent group members from communicating effectively with elected officials in Washington. Among the constraints on grassroots lobbying they identified were the following: misunderstandings about the choices facing lawmakers, contradictory demands for decisions, misperceptions about the group's and/or the representative's position, and avoidance of contact with those legislators who disagreed with the group's policy interests. For the authors, confusion within the interest group constituency appeared to be one of the primary factors in the relatively weak pattern of influence they discovered.

Subsequent research on other organizations suggested that Bauer, Pool, and Dexter's findings were generalizable beyond business groups concerned with tariffs. Rosenau's survey of the membership of Americans for Democratic Action (ADA), for example, revealed that a surprisingly large percentage of group members were unsure about their grasp of policy issues and were not inclined to think that they wielded much influence on specific legislative decisions.[20] Indeed, relatively few of the ADA members responded to the group's call to contact lawmakers in Congress. In addition, Wilson's analysis of the internal communications within several different types of organizations also defined the limits of grassroots activity.[21] The Washington staffs he studied tended to pursue a legislative agenda largely independent of the rank and file. As a result, they could not expect the group's members to share these objectives or to feel intensely enough about them to initiate communication with members of Congress.

The recent surge in interest group organizations and its attendant mobilization of constituents has reawakened interest in the question of the extent and cohesiveness of related political activities of the membership. Much of the current research, however, focuses on Washington representatives and their efforts on Capitol Hill to influence Congress.[22] This approach limits our understanding of the political participation of active citizens (interest group members) in the policy-making process. Nonetheless, some useful insights into the contemporary practice of mobilizing organized constituencies do emerge from this literature, although the evidence from these newly published studies does not pose a genuine challenge to earlier findings.

Loomis, for example, concludes that grassroots lobbying takes so many different forms that it is very difficult even to define as an area of research.[23] Moreover, Godwin's work on environmental groups as well as McFarland's and Rothenberg's analyses of Common Cause indicate that campaigns to mobilize the membership may be mounted to pursue goals of group maintenance, but may also be used to influence Congress.[24] Leadership efforts to rouse the constituency may attract new members, bolster the commitment of existing members, justify the organization's investment in computer technology or stimulate coverage in the news media. These alone would justify the investment of organizational resources in such activities. Finally, it appears that lawmakers and their staffs distinguish between "stimulated" and "spontaneous" communications even though modern mail techniques have made this task more difficult than it was in the past.[25] This practice has the potential of diminishing the impact of the broader political action of interest groups and has prompted groups such as Common Cause to restrict their constituency contact campaigns to a few trusted members in the congressional districts of key committee members.[26]

Although the fragmentary evidence cited above suggests the presence of limits to organized constituent pressure, there are other indications in the literature that appear to justify the emphasis lobbying groups place on such tactics. Basically, the case for membership mobilization rests on the elite character of congressional politics, which gives disproportionate influence to small numbers of intense individuals. The consequences of the elite bias in participation are evident in the attentiveness of House members to their active constituents who contact them on policy issues. Members of Congress spend a disproportionate amount of their time attending to the concerns of this distinctive sector of the congressional district populations.[27] Moreover, the influence of small groups of activists is apparent in several studies of congressional voting behavior. Kau and Rubin, for example, reported a positive relationship between the membership size of environmental organizations in each state and roll call decisions, although their use of state data to analyze House votes limits the applicability of their results.[28] Statistical analyses containing surrogate variables to represent "interests" also predict congressional decisions reasonably well.[29]

Moe's research on decision making of the membership, while not addressing the question of constituency mobilization directly, indicates the presence of two important preconditions for political participation of the membership in the form of grassroots lobbying. First, interest group members must be highly efficacious, with a majority reporting a belief that their contribution to the group matters. Second, a sizable minority of the group's membership must consider politics to be a significant goal in their decision to join an organization.[30] Finally, Fowler and Shaiko, in a Senate roll call analysis, indicate that mobilization of constituent activists has a modest value as a lobbying tool, but they also suggest that certain issues, state conditions, and characteristics of the group activists themselves shape the receptivity of senators to organized constituent opinions.[31]

With these research findings in mind, plausible cases are made *for* and *against* membership mobilization as a (leadership and membership) tool for influencing the government. Contradictory interpretations are possible because group members play a dual role in the business of pressuring policy makers in Washington: they are both potential *advocates* for their organizations and politically active *citizens* within their local constituencies. In either capacity, organizational leaders have limited control over how the members carry out these different roles or how they will be perceived by lawmakers. Nor can group strategists in Washington separate the connections between these relationships; for the context in which membership lobbying campaigns are carried out cannot be divorced from the preexisting network of political relationships between lawmakers and their activist constituents.

When group members are considered political advocates, it is obvious that the decision to participate in group lobbying can be encouraged but not coerced from headquarters. Equally clear is the fact that the membership—not the organization—has the final say over the content of the message sent to Washington when the activity is truly "spontaneous." Organizational leaders may try to control both the participation and the content, but they run the risk of discrediting their members' efforts as "stimulated." Thus, an inevitable constraint on the political participation of public interest group members is the uncertainty over whether or not the members will communicate with elected officials and whether they will reinforce the group's position with reliable signals or undermine it with confusing or contradictory demands.

The political participation of organized constituents therefore encounters some hard realities: The role of *advocate* is a complex one within the organizational context of interest groups, and it is inseparable from the overall political behavior of *citizen activists,* who happen to belong to interest groups. This is why a seemingly simple strategy of influence lends itself to conflicting interpretations and what an analysis of this type of participation must attempt to unravel.

Environmental Organizations, Members, and Communications

It is clear from the preceding analysis of members of environmental groups that belonging to an environmental interest group does not entail the same level of commitment to the organization that one might find in members of religious groups or fraternal organizations, for example (see table 6.1). First, the level of overall self-described group-related activism is quite low with 2 percent (EDF) to 10 percent (SC) of the members considering themselves active in their groups. It follows that the Sierra Club should have the highest level of activism, given the organizational structure of the Club as well as the extensive outing programs (15 percent reported going on a Club outing). However, we might expect a somewhat higher level of activity given the amount of time and resources allocated for this aspect of organizational maintenance. Incidentally, it was the low level of membership participation that caused TWS to discontinue its outing program. Approximately the same levels of activity were reported for other group-related efforts: writing to group leadership and staff, voting for boards of directors, participating in policy discussions, and voting for delegates to national conventions.

In addition to the level of activism, a sizable number of members do not even think of themselves as members; rather they "just send money because [they] think it is a worthy cause." Again, fewer SC members report only a "checkbook affiliation" with their organization. Interestingly, EA and EDF, the smallest of the groups, have the highest levels of membership detachment. This is not to say that the members do not approve of each group's activities. Across the five memberships, 89 percent of the members perceive their group as "representing their own views on environmental and conservation issues to the government." In addition 87 percent believe that their contribution has helped influence the government on such issues, demonstrating the high level of political efficacy among these active citizens. Many members of environmental organization simply do not perceive themselves as integral components in the lobbying process.

Given the heavy reliance on direct mail as a recruiting tool, it is not surprising that members have comparatively weaker attachments to individual organizations than to the larger movement's agenda.[32] Richard Hammond of Names in the News, one of the country's largest mailing list brokers for liberal causes, argued that this pattern of behavior is closely linked to the attributes of the public interest constituency. He simply stated that "the white liberal community in this country is always looking for change."[33] When individuals become a part of the direct-mail network in the United States, the opportunities for change (in the sense that Hammond conveys) are numerous. When it comes time for those renewal checks to be written, many of these citizen activists opt for another group or groups rather than remaining loyal to the same organizations.

For the most part, members choose particular interest groups based on the information available about the groups, the salient issues, and the views of the groups themselves. Judging from the membership responses, the active constituents join these organizations with preconceived opinions and issue agendas. This point places the role of the communication process in a somewhat different light and allows attention to be directed at that aspect of leadership activity and its relationship to the membership participation efforts of the five groups. The vast majority of the group members read the communications sent to them by group leadership; in fact, more than two-thirds of the sample respondents read the environmental magazines and newsletters "frequently."

The magazines and newsletters offered to members of the five groups are diverse in their packaging as well as in their content. The National Wildlife Federation makes a variety of magazines and newsletters available to its members, including: *National Wildlife, International Wildlife, Ranger Rick, Your Big Backyard,* and the *Conservation Report.* However, an associate member receives only *National Wildlife* for the basic membership fee. The other publications may be obtained at an additional cost. The content analysis presented later includes only the *National Wildlife* magazine, for this is the sole publication that all members receive. The *Conservation Report,* a weekly Washington policy update, is an excellent source for keeping informed about the environmental activities in the legislative and executive branches of the federal government. However, only about 2 percent of NWF members receive the publication.

In the 1970s the Sierra Club provided members with ten issues of *Sierra* magazine for the annual membership fee. During the time period of the first content analysis (1977–1978), the Sierra Club changed the format of the magazine slightly as well as its title, from *Sierra Club Bulletin* to the present *Sierra* magazine. Today, Sierra Club members receive *Sierra* bimonthly. Since the late 1970s the magazine has undergone significant changes. Like several of its publishing counterparts, (e.g., *Audubon,* and most recently, *Greenpeace*), *Sierra* is now available for purchase at newsstands and bookstores along with several additions to the environmental journal market—*E, Garbage,* and *Buzzworm.* The Sierra Club has an extensive regional and local organizational structure, and each level of the organization publishes some sort of newsletter for its members. Members may also receive the *Sierra Club National Report,* a national newsletter on group issues, but less than 1 percent of the current members receive this publication. Only *Sierra Club Bulletin* and *Sierra* magazines were included in the content analysis. All members have the opportunity to receive this publication. Nonetheless, in 1977, 4,500 SC members specifically requested that the magazine not be sent to them.[34] Since it is impossible to match the sample respondents to each of the subnational units and since the quality and content of the regional and local newsletters are not consistent across the United States, these publications were not included.

The Wilderness Society published two periodicals in the 1970s. The first, *Living Wilderness* (now simply *Wilderness* magazine), was a quarterly publication very similar in quality and content to *National Wildlife* magazine. In addition to the magazine, TWS published the *Wilderness Report* (monthly, except in June and August), a short (four pages), oversized newsletter including environmental issues and general environmental concerns. TWS ceased publication of *Wilderness Report* in 1982 due to the growing financial burden of the publication. As of April 1996, *Wilderness* has become a biannual magazine, supplemented by a short newsletter.

Environmental Action published *Environmental Action* magazine. In the 1970s members received twenty-four issues for their annual membership fee. Each issue was approximately sixteen pages long and addressed the major environmental issues of the day. By the early 1990s, *Environmental Action* appeared only quarterly, but with an average of forty pages per issue. During the same time period, the Environmental Defense Fund distributed to its members a bimonthly newsletter (four to six pages), the *EDF Letter.* Due to the nature of EDF with its focus on litigation and more technical analyses, there is little time or money available to allocate resources toward improving the communication between EDF leaders and members. Little has changed in the format and frequency of *EDF Letter,* although more recent issues average eight pages.

The Communications Process: Linking Leaders and Members

Joel Thomas, general counsel at NWF, offered the statement quoted at the beginning of this chapter in response to my inquiry about the relative lack of substantive policy content in *National Wildlife* magazine. The National Wildlife Federation is in a unique situation in the environmental movement. It, indeed, has an "undifferentiated mass" of constituents, as compared to other environmental organization memberships. The point Thomas is making, however, is that the internal policy-making process of such large organizations with mass memberships is determined by means other than direct feedback from members. If *National Wildlife* had a higher percentage of policy content, Thomas argued, the leaders would simply hear members' responses to issues presented before them in the magazine pages (i.e., the echoes of the leaders' voices). Therefore, rather than fill their magazine with policy issues (which may or may not cause splits among their members), they have decided to keep the magazine free of policy controversy. Thomas concluded: "We have another policy-making system that does not require us, in effect, to keep them [the members] as apprised [of the organization's issue agenda] or to encourage their interest in specific policy areas."[35]

This strategy may be an effective way of maintaining a membership of almost one million associate members; however, smaller, more activist groups with informed, activist memberships seek out the "echoes" of their members in the external political environment (i.e., through letter-writing, telephone calls, and faxes to public officials). Interestingly, in the responses to survey questions regarding information provided by group leaders, more than one-third of the NWF respondents found the information provided through communications to be personally beneficial. Conversely, less than fifteen percent of the members of EA, EDF, SC, and TWS found the information provided to them by organization leaders to be similarly beneficial. In fact, *National Wildlife* magazine, viewed as a premium, is a significant incentive for members to belong to NWF; two thirds of NWF respondents reported that *National Wildlife* was the "most important reason why [they] belong to the group." Approximately one-quarter of the members of EA, SC, and TWS had similar feelings regarding their publications; virtually no one in the EDF sample felt the *EDF Letter* was at all significant as an incentive for belonging to the organization.

While communications with members clearly serve a variety of purposes for the organizations' leaders, including a significant organizational maintenance function, many leaders seek to use these communications as representational tools—to present the salient environmental issues of the day and to inform members as to the institutional focus of those issues. In 1977 each of the groups solicited members as interest group *advocates* to take part in leadership-administered membership mobilizations through issue alerts and special mailing and telephone lists. However, those partic-ipating in the leadership strategy represent only a small portion of the total membership involved as politically active *individuals* who write letters to public officials on environmental matters without direct interaction with group leaders.

These active individuals are the targets of the policy-laden communica-tions sent by organization leaders. Earlier it was noted that the level of membership activity in the group organizations was quite low and that a sizable number of group members did not regard themselves as members; they simply felt that the group was working for a good cause. Nonetheless, the majority of those members *not* perceiving themselves as group mem-bers were active individuals, participating in the policy-making process by writing letters to members of Congress. (More than two-thirds of EA and EDF "nonmembers" contacted their representatives in Washington.) Sim-ilarly, less than 10 percent of all members consider themselves "active" in group efforts, yet between 46 percent and 74 percent of the members of the five organizations have individually sought to influence policy outcomes by contacting public officials.

One pattern emerges—members of environmental groups do not con-sider the act of writing or contacting a member of Congress or the presi-dent or the secretary of the interior about an environmental issue under de-bate to be an interest group activity. Equally evident is the fact that these members are writing letters regularly and in large numbers. This form of political participation, like the act of joining an interest group, may be a positive reinforcement for lobbying activity conducted on Capitol Hill by interest group leaders. But we should not assume that environmentalists, both within groups and between groups, share a single point of view on a wide range of environmental issues.[36] Group leaders, however, do have at their disposal the means of keeping members aware of the salient issues and of directing members to the proper outlets for registering their con-cerns. It is within this context that we will analyze the communications be-tween leaders and members of these five environmental organizations and their possible impact on the issue agendas of members.

Content of Environmental Communications in the 1970s

National Wildlife Federation

NWF members received a very attractive magazine, *National Wildlife*, when they joined the group. Every other month, the publication of approx-imately sixty pages was sent to members. Clearly the dominant feature of the magazine was its excellent full-color nature and wilderness photography as well as its skillfully created drawings, paintings, and illustrations. Together, the pictorial essays and illustrations accounted for over half of the total magazine content for the year. Group advertisements accounted for a little less than 10 percent of the content. NWF made holiday shopping quite simple with the range of products offered for sale. From books, calendars, prints, and posters to clocks, binoculars, neckties, and glasses, NWF offered its members discount prices on a wide variety of items.

Upon my request for information about the group and a membership application, I was bombarded with NWF book and merchandise catalogues. Not once were the environmental goals of the group mentioned in the proselytizing literature, only the "great benefits" one would receive for joining: *National Wildlife,* "20% discount on selected nature books and records, vacation opportunities in U.S. and Canada, travel opportunities in our SAFARI program, personalized membership card, colorful window decal, and full-color wildlife stamps." Compared with the recruiting letters of the other four groups, the NWF literature compares favorably with that of the "Book-of-the-Month Club." By the early 1990s, NWF merchandise sales amounted to more than $65 million annually.[37]

Roughly 30 percent of the magazine was comprised of brief notes and longer articles (see table 6.2, column 1). However, the majority of these notes and articles had no policy content. *National Wildlife* has always been an excellent magazine for someone who wants to learn about building a garden, hiking, bird watching, natural dyeing of clothes, fishing, and even hunting. The vast majority of the essays and articles focused on nonpolicy issues such as these. With all of the photography, illustrations, advertisements, and nonpolicy notes and essays, only about 8 percent of the content remained for the presentation of policy issues.[38]

The National Wildlife Federation provided its members with the least amount of general and specific policy content. Even the *EDF Letter,* with a total of only thirty pages for the entire year, mentioned more specific policy issues than *National Wildlife,* with more than 400 pages in all.

NWF editors did present twenty-eight specific policy issues, mostly in the form of brief notes appearing in a special section of each issue entitled, "Washington Scene." In only two cases were the readers asked to take action on legislative matters; in seven cases there were requests for action on executive agency issues (see table 6.3). In summary, *National Wildlife* is an attractive, colorful, easy-to-read, family-oriented magazine—perfect for the coffee table or doctor's office. However, as an information mechanism for active individuals, it is comparatively useless.

TABLE 6.2 Content Analysis: Environmental Group Communications, 1977–1978

	NWF	SC	TWS	EA	EDF
Advertisements	8.4%	24.8%	0.3%	3.2%	6.3%
Photography	40.2	13.9	30.5	11.0	3.9
Letters to the Editor	0.2	1.8	1.0	5.3	0.0
Illustrations	11.7	4.2	3.3	6.6	1.8
Book Reviews	0.0	4.2	2.0	4.1	0.0
Poetry	0.0	0.5	0.2	0.0	0.0
Group News, Activities	4.6	12.3	5.2	6.6	21.1
Brief Items and Notes[a]					
Non-Policy Issues	1.7	0.0	0.3	0.2	0.0
General Policy Issues	1.6	1.0	2.0	2.2	8.7
Specific Policy Issues	1.5	1.8	6.0	19.2	40.3
Articles/Essays[b]					
Non-Policy Issues	24.7	15.5	21.0	4.6	0.0
General Policy Issues	4.6	12.0	18.3	23.9	12.5
Specific Policy Issues	0.7	7.8	9.9	13.2	5.6
	99.9%	99.8%	100.0%	100.1%	100.1%
Total Pages	436	501	304	386	30
Total Issues	7	10	16[c]	24	7
Overall Policy Content	8.4%	22.6%	36.2%	58.5%	67.0%
Specific Policy Content	2.2%	9.7%	15.9%	32.4%	45.9%

[a]Brief Items and Notes = items of one-half page or less in length.
[b]Articles/Essays = items longer than one-half page.
[c]Number of Issues includes eleven *Wilderness Reports* and five *Wilderness* magazines.

Sierra Club

The Sierra Club communications were quite similar to those of the NWF, although there are some important differences. The most noticeable difference between *Sierra Club Bulletin/Sierra* and *National Wildlife* is quite apparent as one pages through each issue. SC sells advertising space to outside commercial enterprises (DuPont, Vibram Shoes, Gerry Camping Gear, L. L. Bean, etc.). This is important given the costs of publishing a full-color, forty-eight-page magazine nine times a year.[39] However, the advertisements account for one-quarter of the magazine content. Photography and illustrations (including cartoons) account for roughly 20 percent more. *Sierra Club Bulletin* also contains reviews of recent publications of potential interest to the members. One also finds some poetry interspersed with nature illustrations. Another substantial portion of the magazine is composed of interest group news, activities, and information. With its extensive outings program, SC must publish information regarding the many overseas trips as well as the wilderness outings held throughout the United States (see table 6.2).

The aforementioned items account for approximately two-thirds of the

TABLE 6.3 Content Analysis: Types of Policy Issues, 1977–1978

	NUMBER OF BRIEF ITEMS & ARTICLES	INSTITUTIONAL FOCUS		
		Legislative	*Executive*	*Judicial*
NATIONAL WILDLIFE FEDERATION				
Non-Policy Issues	88			
General Policy Issues	34			
Specific Policy Issues	28	13	12	3
		(2)[a]	(7)	(0)
THE WILDERNESS SOCIETY				
Non-Policy Issues	18			
General Policy Issues	39			
Specific Policy Issues	91	60	26	6
		(31)	(15)	(0)
SIERRA CLUB				
Non-Policy Issues	45			
General Policy Issues	56			
Specific Policy Issues	63	41	18	4
		(12)	(1)	(0)
ENVIRONMENTAL ACTION				
Non-Policy Issues	12			
General Policy Issues	79			
Specific Policy Issues	244	139	86	19
		(70)	(6)	(0)
ENVIRONMENTAL DEFENSE FUND				
Non-Policy Issues	0			
General Policy Issues	16			
Specific Policy Issues	50	8	22	20
		(3)	(1)	(0)

[a]Number in parentheses indicates the number of instances where members were specifically instructed to take political action (i.e., write a letter or send a telegram to a federal official).

total content. The remainder of the magazine content is left for brief notes and essays and articles. Nonpolicy articles and notes account for 15 percent of the content. As with *National Wildlife,* the nonpolicy essays occupy greater amounts of space than specific or general policy articles, although *Sierra* contains more policy-related articles than nonpolicy ones (see tables 6.2 and 6.3). Many of the brief notes appear in the "News" section of each issue (roughly two full pages). About forty specific legislative issues were addressed with twelve requests for membership action. Half as many executive branch issues were mentioned with only one request. Four court cases were presented as well. Overall, *Sierra Club Bulletin/Sierra* fares better than *National Wildlife* as an information mechanism for grassroots lobbying of the membership. Still, only 10 percent of the total content concerns specific policy issues (see table 6.2).

The Wilderness Society

TWS sent members two publications in 1977–1978: *Living Wilderness* and the *Wilderness Report.* These communications were quite different in packaging and content. *Living Wilderness,* like the two previously mentioned magazines, is colorful and attractive. With almost no group advertisement, one-third of the magazine is filled with excellent wildlife photography. Illustrations, book reviews, letters to the editor, poetry, and interest group news account for an additional 10 percent. There are no brief items or notes; each issue contains five to seven long essays on policy and nonpolicy issues. Of the remaining 55 percent, nonpolicy essays make up 25 percent. Again, one reads about wilderness hiking, trips to exotic lands, and bird tracking. Policy-related essays make up the remaining 30 percent. Thirteen general policy issues were mentioned, while eight articles focused on specific legislative concerns (two of which requested membership action). One article is directed at the Carter administration.[40]

While the *Living Wilderness* magazine addresses some environmental policy issues, the vast majority of the TWS environmental agenda is set in the *Wilderness Report.* With ten issues a year, TWS is able to track environmental issues through the entire policy process. Each issue, a four-page, 11″ × 15″ newsletter, includes some black-and-white photography, illustrations, and interest group information, accounting for roughly one-third of the total content. The remainder of the content is related to policy. With limited space, most policy issues are presented as brief notes. Specific policy issues dominate the items presented, with the total number of issues presented surpassed only by *Environmental Action,* discussed below. In addition to the larger number of policy issues, the number of requests for membership action is equally impressive. Together, *Living Wilderness* and the *Wilderness Report* offer an excellent mix of glossy packaging and environmental issues.

Environmental Action

Leaders of Environmental Action did a superb job in setting an environmental agenda for their members in 1977 and 1978. With comparatively little content composed of (black-and-white) photography, illustrations, advertisements, letters to the editor, book reviews, and interest group news, the vast majority of the magazine content (twenty-four issues/year, sixteen pages/issue) is policy related (see table 6.2). The number of specific (and general) policy issues presented (as well as the number of requests for membership action) is astounding. In addition to the essays and articles, *Environmental Action* has two sections, "Econotes" and "On the Hill," in which most of the policy content is included. In the "On the Hill" section alone, over sixty specific issues are presented, most of which are followed by a "What to Do" note—instructing members as to their most effective action on a particular issue (see table 6.3).

With almost 250 mentions of specific policy issues and more than seventy-five mobilization efforts by EA leaders, *Environmental Action* magazine provides by far the most extensive environmental agenda of the five organizations. Further, the specific policy guidance and direction the editors offer members is far superior to that offered by NWF, SC, and EDF. Even the excellent coverage by TWS in the *Wilderness Report* does not match the breadth of coverage presented in *Environmental Action* magazine.

Environmental Defense Fund

The EDF did not spend a great amount of its limited resources on the publication of the *EDF Letter* in 1977. The newsletter is created in EDF's New York City headquarters, which receives input from its regional offices. Though the group published only thirty pages in 1977–1978, most of the newsletter content is related to policy. In fact, there are no non-policy issues presented. Almost half the content is focused on specific policy issues (see table 6.2). There is a disproportionate amount of coverage of judicial action; this follows from the focus of the organization. Very few legislative issues are presented, while more than twenty issues involved executive agencies and the president. In only three cases were members requested to take action on specific issues (see table 6.3).

Even with its short newsletters, EDF manages to present a fair amount of environmental policy issues. While coverage is not as extensive as that of

EA and TWS, it is better than that of NWF and is close to that of SC. The weakness of the EDF coverage lies in its inability to direct members in their individual efforts at contacting public officials on specific policy issues.[41]

Before analyzing the changes that have taken place in the communications between environmental leaders and their members over the past twenty years, a crude assessment of the impact of these earlier communications will be provided with the help of the membership data presented in chapter 5.

Environmental Agendas and Issue Salience

In my earlier research the opinions and intensities of opinions of environmental group members on numerous issues were analyzed. Not only were there differences in the opinions of members on major policy issues, but the intensities of these opinions were also variable, both within groups and, even more so, between them.[42] In this section, these two attitudinal components will be addressed. As previously stated, these members join environmental organizations with preconceived opinions on environmental issues. Given the transience of environmental group members (average tenure is less than three years) and the relative abundance of organizational alternatives, it is likely that members will exit groups with inconsistent viewpoints rather than modify their own opinions. Interest group communications may, however, raise the salience levels of issues that are consistent with members' views, particularly if the issues stress potential environmental threats and are presented in an emotional manner. For example, if a member of EA believes that nuclear power is a viable energy alternative, it is more than likely that he will not be swayed by the articles in *Environmental Action*. However, a "no-nuker" in EA, already attentive to the issue, will selectively focus on nuclear power articles and come away with a greater potential to take action (e.g., letter-writing, contacting other groups more directly focused on the issue, and even protesting).

Another point that must be addressed prior to analyzing issue salience is the intervening factor of multiple membership. The percentages of members belonging to the other groups are reported in table 6.4 and taken into account in the concluding analysis. The data from members of all five organizations reveal some interesting patterns regarding multiple memberships. When analyzing only respondents sampled from the 620,000 NWF associates, one finds that only 2 percent of the members belong to EA and EDF; in fact, 70 percent of NWF members had never heard of these groups. As the table shows, 6 percent belong to SC and 3 percent to TWS.

TABLE 6.4 **Membership Statistics**

	NWF	TWS	SC	EDF	EA
% Members of group for 1 year or less	23	15	21	18	33
Median length of membership (years)	3	4	3	3	2
Sample N	509	623	661	630	705
Multiple memberships (N reporting membership in each org.)	1289	1110	1347	897	1100

Yet, when the data are aggregated, over 40 percent of the 3,128 respondents claim membership in the National Wildlife Federation. Further, NWF respondents belong to an average of *1.4* groups; whereas EA and EDF members, for example, average about *6* organizational affiliations. In this sense, the aggregate sample is biased in the direction of the more activist components of the national environmental movement. Some respondents who were members of several groups and were sampled from EA and EDF membership lists belong to NWF, but it is evident that these members are not at all representative of the larger NWF constituency.

Regarding the issue agendas of the five organizations, the results in table 6.5 suggest that a general wilderness agenda exists within the group memberships.[43] The general environmental issues for NWF are near the top of the list with water pollution one of the salient issues. Few NWF members, however, feel that it should be the top priority of the group. The major priority is wilderness preservation, yet only 12.8 percent find it to be the most important issue. Even among the top ranking issues, barely a majority of the members feels intensely about the issues. Note that opposition to nuclear power is sixteenth.

Wildlife management for improved fishing and hunting ranks higher (thirteenth) than the nuclear power issue. This is not surprising, though. NWF is primarily a conservation group with a comparatively more conservative point of view. It would not be uncommon for NWF members to belong to the National Rifle Association. On the agendas of the remaining four groups, wildlife management for improved fishing and hunting ranks last or second to last. Generally, the issue agenda of NWF members is comparatively loosely organized; no one issue dominates and there is little consistency throughout.

TABLE 6.5 Issue Salience Among Members of Five Environmental/
Conservation Organizations

Issue	NWF	SC	TWS	EA	EDF
Water Pollution	1[a]	2	3	2	1
Air Pollution	4 (7.9)[b]	3 (6.8)	5 (6.4)	1	2 (9.7)
Wilderness Preservation	5 (12.8)	1 (25.9)	1 (24.8)	4 (9.8)	5
Preserving Natural Areas	3 (9.3)	4 (6.5)	2 (7.3)	5	4
Alternate Energy Sources	6 (11.9)	5 (10.6)	6 (8.7)	3 (14.3)	3 (11.9)
Wildlife: Animals/Birds	2 (10.0)	6	4	10	7
Toxic Substances	8	7	7	6 (6.9)	6 (9.7)
Animal Protection	7	8	8	12	11
Rational Planning	9	10	9	9	9
Population Problems	11	9 (10.8)	10 (11.0)	8 (12.4)	8 (15.5)
Opposition/Nuclear Power	16	12	14	7 (15.6)	10 (11.0)
Food Supply	10	14	12	16	15

[a]Rank ordering of the top twelve issues for all five groups is based on member responses to a list of twenty-two policy issues. Each respondent was asked to identify the issues that were "very important to you personally"; in addition, they were to select the issue they considered to be the "most important" to them from the list. For example, most NWF members chose Water Pollution as a "very important" issue (63.0%); SC members most often indicated that Wilderness Preservation was "very important" issue to them personally (67.4%). TWS members chose the same issue most often (72.8%). EA members chose Air Pollution (66.2%) and EDF members selected Water Pollution most often (69.2%).

[b]Percentages in parentheses indicate the members who chose each issue as "the most important" issue to them personally. Top five "most important" issues are presented for each group.

Conversely, SC members demonstrate slightly more cohesiveness regarding issues. Approximately 7 percent of the members belong to EA and EDF and about 20 percent belong to NWF and/or TWS; yet when the data are aggregated, almost half of all respondents belong to the Sierra Club. Wilderness preservation is the most salient issue among SC members, with over two-thirds believing that the issue is "very important" personally and over one-quarter ranking it as the "most important" issue. Beyond the wilderness issue, SC members also feel strongly about energy conservation, population problems, air and water pollution, and toxic substances. Opposition to nuclear power remained less salient, even though their opinions on the issue were decidedly antinuclear. One finds a policy agenda focused on wilderness issues as well as several other environmental issues.

The Wilderness Society's agenda is as one would expect. The dominant issue area is wilderness preservation. But like SC members, TWS members also consider population problems, air and water pollution, alternative energy sources, and energy conservation to be important issues. TWS has a high incidence of multiple membership, with 10 percent of its members claiming affiliation with EA, 24 percent with EDF, 29 percent with NWF, and 37 percent with SC.

Members of Environmental Action have the most diverse issue agenda of the five groups. While air and water pollution are dominant concerns, one also finds alternative energy sources, wilderness and natural area preservation, toxic substances, and opposition to nuclear power to be issues of some concern to EA members. Note the breadth of the most important issues. Four quite distinct environmental issues are intensely maintained by blocs of members within the group (ranging from 10 to 15 percent), whereas the other groups (SC, TWS, and NWF) have only one dominant issue or none at all.

Members belonging to more than one organization make up the majority of the memberships of both EA and EDF. Over 90 percent of the groups' respondents belong to at least one of the remaining four groups. One half of EA members belong to EDF; one-third belong to SC, TWS, and/or NWF. Similarly, one-quarter of EDF members belong to EA, and one-third also belong to SC, TWS, and/or NWF. The EDF issue agenda resembles that of EA members, equally as diverse, with similar intensities across issues. Population problems arise as the most important issue; yet just over half feel that it is an important environmental issue.[44]

These findings point to a variety of relationships between agendas presented in communications from the leadership and relevance of the issues for the membership. In the cases of SC, TWS, NWF, and EA, the direct connections are more apparent. EDF presents somewhat of an anomalous case. One plausible explanation for the cohesiveness and diversity of the issue agenda of EDF members is the impact of multiple membership. Perhaps the multiple sources of issue content provide EDF members with sufficient information to maintain a coherent issue agenda and to maintain a very high level of membership activity (letter-writing). One must not discount other sources of environmental policy content either. While these communications are focused on environmental policies, they are distributed relatively infrequently, with the exception of TWS and EA publications.

Daily newspapers, especially the major national publications, and weekly magazines (especially the more policy-oriented weeklies published in

Washington, e.g., *Congressional Quarterly* and *National Journal*) keep tabs on environmental matters debated in Washington. The members of these organizations are highly educated professionals, able to derive information from numerous sources. They are certainly not limited to membership-based environmental magazines. A growing magazine market focused on environmentalism provides this attentive public with additional sources of information about the environment. They are limited, though, to interest group publications for direct or indirect cues from group leaders about the important issues facing the group and about the relevant avenues through which members may voice their opinions. In most cases newspapers do not provide these services. While environmental groups do not have the budgets of Mobil or Exxon to buy full-page advertisements in the *New York Times* with any frequency, these organizations as well as other public interest and single-issue groups have attempted to "go public" with their message through the print media and television to disseminate their information.

As a result, the direct contacts with individual members may be less important in a wider public communication strategy, including the growing use of the internet, which will be discussed in the concluding section of this chapter.

Communications and Membership Political Participation

While no sweeping generalizations may be made about the impact of communications from the interest group leadership on the membership's political activity, there are several conclusions that may be drawn from the preceding analysis. It is important to begin with what was known at the outset. First, environmentalists write public officials about environmental matters; some are directly stimulated by group leaders through issue alerts and mailing lists, but a much larger percentage of members are not driven by direct mobilization. Second, political activity (citizen participation) is linked to educational attainment, income, race, and age.[45] Within these public interest groups are joined society's politically active individuals. They are white (97 percent), upper-middle class, and highly educated—i.e., predisposed to political action. This predisposition must be introduced as an intervening factor. One could easily compare membership activity in NWF with the communication content and conclude that the lack of indirect cuing by group leaders has resulted in an inactive membership. This, in fact, may be the case. However, the composition of NWF causes one to consider the possibility that membership in NWF is not necessarily perceived by all members as a political activity, and hence further activity may not readily follow.

In spite of this point, it is sufficiently evident that the communications within interest groups serve to enhance the group members' awareness of issues and may translate into an effective cuing mechanism for their activity. It is unfortunate that the data do not allow direct analysis of the content of member communications with public officials. Nevertheless, the use of opinion and intensity data do give some indication of the range of possible issues addressed by active individuals in environmental groups. There exists an identifiable pattern between the issue agendas set by interest group leaders and the significance of issues for group members. It is not a major theoretical leap to link the latter with possible issues presented by the members to elected officials in Washington.

Table 6.6 provides some indication of the degree of cohesiveness regarding issues among the more active elements of environmental organizations. The results presented in table 6.6 include only the responses of members who have written to policy makers regarding environmental matters. As the opinion questions presented include dimensions of agreement and disagreement, it is possible to evaluate both the opinion positions of these active members as well as the intensities of their beliefs.[46]

As all mean scores reported are positive, there is at least a general issue consensus among active members of environmental organizations. However, there are some scores that deserve closer scrutiny. For example, the NWF "Opposition to Nuclear Power" approaches zero. Obviously, the active membership of the National Wildlife Federation is split on this particular issue. The leadership of NWF does take a position on nuclear power (antinuclear), but one would have a hard time finding evidence of the position in the leaders' communications to their members.

In comparing issue responses across organizational memberships, one finds that the issue of runaway technology garners the least support among these activists. Conversely, population problems, wilderness preservation, and energy conservation appear to be issues around which environmental leaders could most easily mobilize their forces. Overall, it is evident, nonetheless, that even the active members of NWF are the least cohesive of environmental constituent activists. The mean scores in table 6.6 for each group are: NWF = 0.859; TWS = 1.120; SC = 1.100; EDF = 1.216; and EA = 1.213. Difference of means tests indicate that the NWF mean responses are significantly different (lower) than the responses of the other four organization's memberships. Of the remaining active memberships, none offers any particular reason why their leaders should be overly con-

TABLE 6.6 Opinions of "Active" Environmental Organization Members, 1977–1978[a]

ISSUE	NWF (N = 209)	TWS (N = 395)	SC (N = 353)	EDF (N = 379)	EA (N = 508)
Runaway Technology	0.139[b]	0.294	0.150	0.503	0.553
Wilderness Preservation	1.197	1.559	1.491	1.374	1.388
Alternative Technology	1.131	1.233	1.104	1.234	1.274
Population Problems	1.337	1.446	1.481	1.467	1.326
Opposition to Nuclear Power	0.010	0.703	0.893	1.317	1.467
Animal Protection	1.177	1.355	1.236	1.219	1.236
Air Pollution	0.655	0.956	1.006	1.124	0.984
Water Pollution	1.036	1.003	0.880	1.033	1.023
Energy Conservation	1.273	1.131	1.637	1.570	1.533
Toxic Substances in Food/Water	0.630	1.131	1.118	1.323	1.343

[a] Samples include only those respondents who had contacted public officials on environmental matters during the year prior to the survey.

[b] Mean opinion scores were calculated for the survey responses to the questions. The responses were coded 2, 1, −1, −2, from strongly agree (very serious) to strongly disagree (not serious). Among the respondents, there was general agreement in direction of response. However, the scores reflect the differences in range of responses. A mean opinion score of 2.0 indicates that the entire "active" membership in the organization is in total positive agreement on the issue; a score of −2.0 would reflect a total negative agreement. A score approaching zero would indicate a significant split in opinion among members on the particular issue.

cerned with the letter-writing activities of their constituent-activist members.

In general, leaders of Environmental Action and The Wilderness Society were quite effective in establishing clear issue agendas for their members and in providing ample cues for membership action. Over 60 percent of their members contacted officials regarding environmental matters. Given the findings in table 6.6, some effectiveness may be attributed to interest group communications in the process of indirectly mobilizing active individuals within these groups. The relationship between communications and activity of the interest group members is less clear at the extremes. The dilemma of attributing ineffectiveness to NWF leaders in failing to set an adequate issue agenda has been discussed. One must remember that the goal of NWF publications is not political mobilization. EDF, at the other extreme, provides little usable policy content to its members; yet almost three-quarters of its members actively voice their opinions to policy makers in Washington, and in a fairly cohesive manner as well.

Members belonging to several groups simultaneously may explain the importance of issues for the membership and the members' activity partially; the demographic composition of the membership may also explain its political activity. One final plausible explanation may be linked to the issues themselves. Environmental organizations may try to "manufacture" issues, but the agenda is largely controlled by the institutional actors in government, though lobbyists do infiltrate the policy process. In the year of the content analysis, there were only a few instances where leaders sought the efforts of members to bring an issue onto the legislative agenda. For the most part, members are involved after a bill is introduced. EDF members may have been disproportionately attentive to the issues that reached the 95th Congress, which may have caused them to be comparatively more active.

In summary, the content analysis and membership assessment for the 1977–1978 period yield the finding that interest group communications are potentially quite effective at setting a coherent environmental agenda. Further, they serve to enhance the possibility of indirect political participation of the membership by drawing members' attention to the important issues under consideration and by directing them to the proper destinations of their communications with public officials. However, the communication processes between leaders and members of the five organizations have undergone significant changes during the last two decades. Today, the messages to members appear less often than in the past. In all five cases, the frequency of distribution of magazines and newsletters has declined. Furthermore, in four of the five organizations the policy content of the communications has diminished significantly. In the content analysis that follows, the coding schemes utilized in the analysis of the 1977–1978 communications are again applied to the communications of the organizations for a one-year period during 1991–1992. These communications are representative of the contemporary links between organization leaders and their members.

Content of Environmental Communications in the 1990s

National Wildlife Federation

The National Wildlife Federation communications offered to all members remain packaged in their magazine, *National Wildlife*. Today members receive six issues per year, one less than in 1977–1978.[47] While the length of each magazine has not changed dramatically, the specific policy content of the communications has increased almost sixfold. The content analyses for 1991–1992 are presented in table 6.7. The NWF content analysis appears in column 1; below the 1991–1992 analysis, the overall figures for pages, issues, and content are reprinted from table 6.2 for purposes of comparison. Aside from written content, the major changes involve an increase in photography with a corresponding decrease in advertisements and illustrations.

In 1977–1978 *National Wildlife* was virtually devoid of any specific discussion of policy issues. In the 1990s almost a quarter of the magazine is filled with policy-related content and 12 percent deals with specific policy issues, up from only 2 percent in the earlier analysis. While this level of specific policy content may not be enormous, one must remember that NWF remains the largest of the five organizations. What explains the increase in policy information disseminated from leaders to members? The most obvious explanation lies in the differences in editorial philosophy between a father and his son.

For almost thirty years, John Strohm guided the growth of the magazine and controlled its composition. His death in 1987 caused a significant break in the continuity of the communications business of the Federation. He was succeeded by his son, Robert, who has worked on the magazine since 1968. Today, with editorial reins held by Robert Strohm, the magazine reveals a new philosophy and management strategy. While it remains attractive and entertaining, it contains much less of the nonpolicy "pap," to borrow from NWF board member John Gottschalk (eighty-eight stories in 1977–1978 versus twenty-eight in 1991–1992), and more policy-related stories, including items about issues such as wildlife conservation, energy use, water pollution, forests, wetlands, and toxics (see table 6.8).

TABLE 6.7 Content Analysis: Environmental Group Communications, 1991–1992

	NWF	SC	TWS	EA	EDF
Advertisements	3.7%	35.8%	5.0%	7.0%	1.7%
Photography	51.0	18.7	26.2	14.4	16.1
Letters to the Editor	0.0	2.4	3.7	1.9	0.0
Illustrations	6.1	4.4	4.5	5.0	2.1
Book Reviews	0.0	2.0	5.3	2.0	0.6
Poetry	0.0	0.0	1.3	0.0	0.0
Group News, Activities	1.6	4.2	14.4	8.4	10.6
Brief Items and Notes[1]					
Non-Policy Issues	3.5	0.0	0.9	0.6	1.2
General Policy Issues	0.0	0.2	0.6	5.1	8.9
Specific Policy Issues	7.2	0.2	3.1	4.5	14.5
Articles/Essays[2]					
Non-Policy Issues	10.8	23.4	12.8	1.4	10.1
General Policy Issues	11.0	6.2	20.4	23.5	17.3
Specific Policy Issues	5.1	2.5	1.9	26.1	17.3
	100.0%	100.0%	100.1%	100.3%	100.0%
Total Pages	344	778	144	180	40
Total Issues	6	6	4	5	5
Overall Policy Content	23.3%	9.1%	26.0%	59.2%	57.6%
Specific Policy Content	12.3%	2.7%	5.0%	30.6%	31.8%
		1977–1978			
Total Pages	436	501	304	386	30
Total Issues	7	10	16	24	7
Overall Policy Content	8.4%	22.6%	36.2%	58.5%	67.0%
Specific Policy Content	2.2%	9.7%	15.9%	32.4%	45.9%

[1]Brief Items and Notes = items of one-half page or less in length.
[2]Articles/Essays = items longer than one-half page.

According to Robert Strohm, a conscious decision was made in late 1990 to shift the focus of the magazine toward substantive issues, specifically those related to nature, sciences, and the environment. Of the three components, the latter was stressed during the time under analysis. According to Strohm, the editorial board had a "gut feeling that members were more interested in contemporary issues relating to public policy." Interestingly, Strohm felt that he may have overcompensated and is attempting to rebalance the policy mix with fewer specific environmental issues presented. Nonetheless, Strohm argues that the overall pattern of increased policy content will remain: "readers are more interested; more is at stake today."[48]

Sierra Club

From a marketing perspective, *Sierra* magazine has experienced the most dramatic change of any of the communications analyzed. *Sierra* is no longer solely a membership-oriented magazine. Today, anyone may purchase *Sierra* as it is available at newsstands and at the major chain bookstores across the country. This broad availability constrains the editors of the magazine. Essentially, the magazine remains an instrument of the Sierra Club; as such, its editors are responsible for some aspect of maintaining the membership base. Efforts at maintaining membership or informing members of club or policy activities may or may not coincide with efforts made toward gaining a larger readership outside the organization. As a result there is tension between editorial staff and program staff regarding the content of the magazine.

From the evidence presented in tables 6.7 and 6.8, it seems as though the efforts at expanding the market have led to a weakened policy agenda as the overall policy content has dropped from almost a quarter of the magazine to less than 10 percent. The frequency of distribution has also dropped from ten issues annually to six bimonthly issues. Each issue is much larger today than it was fifteen years ago (130 pages compared to 50 pages, on average); however, much of this growth is explained by increased advertising space. More than a third of the magazine is filled with advertisements (278 of 778 pages), by far the highest percentage of all the communications analyzed.

Regarding the specific policy content of *Sierra*, today readers receive only about a third as many specific policy articles (twenty-two in 1991–1992 versus sixty-three in 1977–1978) although the number of mobilization efforts is roughly equal in both periods. The 1991–1992 issue agenda includes articles dealing with ANWR, ancients forests, National Parks, GATT, spotted owl, NAFTA, asbestos, biodiversity, nuclear testing, endangered species, wetlands, and the Earth Summit. While *Sierra* and its predecessor, the *Sierra Club Bulletin,* have included outside advertising as an important component in their content, today's *Sierra* represents a marked alternative in the interest group communication process. Distinct from its counterparts, *Sierra* incorporates the production values found in the major com-

TABLE 6.8 Content Analysis: Types of Policy Issues, 1991–1992

	NUMBER OF BRIEF ITEMS & ARTICLES	INSTITUTIONAL FOCUS		
		Legislative	Executive	Judicial
NATIONAL WILDLIFE FEDERATION				
Non-Policy Issues	28			
General Policy Issues	42			
Specific Policy Issues	34	12	21	1
		(2)[a]	(2)	(0)
Specific Policy Issues	28	13	12	3
(NWF, 1977–1978)		(2)	(7)	(0)
SIERRA CLUB				
Non-Policy Issues	74			
General Policy Issues	31			
Specific Policy Issues	22	13	9	0
		(7)	(4)	(0)
Specific Policy Issues	63	41	18	4
(SC, 1977–1978)		(12)	(1)	(0)
THE WILDERNESS SOCIETY				
Non-Policy Issues	15			
General Policy Issues	11			
Specific Policy Issues	23	17	5	1
		(2)	(0)	(0)
Specific Policy Issues	91	60	26	6
(TWS, 1977–1978)		(31)	(15)	(0)
ENVIRONMENTAL ACTION				
Non-Policy Issues	10			
General Policy Issues	41			
Specific Policy Issues	69	28	36	5
		(7)	(2)	(0)
Specific Policy Issues	244	139	86	19
(SC, 1977–1978)	(70)		(6)	(0)
ENVIRONMENTAL DEFENSE FUND				
Non-Policy Issues	7			
General Policy Issues	20			
Specific Policy Issues	31	9	14	8
		(3)	(3)	(0)
Specific Policy Issues	50	8	22	20
(EDF, 1977–1978)		(3)	(1)	(0)

[a]Number in parentheses indicates the number of instances where members were specifically instructed to take political action.

mercial magazines. This adoption of a unique dual-marketing strategy is not without consequences, however.

The Wilderness Society

The most significant change in the communication process of the Wilderness Society lies in the frequency of distribution. In 1977–1978, TWS members received monthly issues of *Wilderness Report* in addition to their quarterly magazine. Today, only the biannual magazine, *Wilderness*, links leadership and membership, along with a short newsletter published periodically.[49] Not only has the frequency declined but the volume of content as well. In 1977–1978 TWS members received 304 pages of information, whereas current members received only 144 pages in the 1991–1992 period. Therefore, while the decline in policy content is evident in table 6.7, the more dramatic difference lies in the number of issues communicated, as presented in table 6.8. In 1977–1978, ninety-one specific policy issues were presented to the members with forty-six efforts at mobilizing them to action. In the more recent analysis, only twenty-three issues were presented with only two mobilization efforts.

This precipitous decline is primarily the result of publishing the magazine less frequently. With publisher deadlines falling often two months prior to distribution to members, it is difficult, if not impossible, to provide relevant coverage of ongoing policy issues. Furthermore, attempts to mobilize are made difficult by the unpredictability of the policy-making process in Congress and in the executive branch. As a result the magazine has taken on a retrospective, rather than prospective, approach to communicating policy issues.

A secondary reason for the decline in policy content stems from the difficulty that arose in the mid-1980s between the Wilderness Society and the Internal Revenue Service (see appendix D). Directly or indirectly, the IRS action has had a chilling effect on the propensity of editors to delve into specific legislative battles. This pattern is evident in the content of magazines published in 1991–1992 as a limited wilderness agenda is presented, including an ANWR update, abuses in the Fish and Wildlife Service, endangered wolves, forest protection, California Desert Protection Act, wilderness management, grazing fees, and endangered salmon.

Environmental Action

Like the Wilderness Society, Environmental Action also truncated its messages to members. Interestingly, while the policy content of the communications has remained constant, the ability to deliver these policy issues to the membership diminished significantly; members receive less than half the total amount of pages they received fifteen years ago. And, more important, the frequency of distribution has virtually negated the mobilizing efforts of the past. Rather than receiving an issue every other week, members more recently received their communications at most every other month. As a result, *Environmental Action* included coverage of 69 specific policy issues with nine accompanying mobilization efforts in 1991–1992, whereas the magazine contained 244 specific issues with seventy-six mobilization efforts fifteen years earlier.

This is not to say the Environmental Action staff did not maintain and present a broad policy agenda for the members. To the contrary, by contemporary standards the issue agenda, as communicated to members, remained the most inclusive of any of the environmental organizations. Nonetheless, the agenda presented in 1991–192 was little more than a quarter the size of the organization's issue agenda in 1977–1978. Environmental Action, perhaps more than any other group, has consistently felt the economic pinch. The costs associated with the production and publication of twenty-four issues annually became unbearable. As a result, Environmental Action was forced to operate within its limited means. Given the ever-increasing costs of production and distribution, coupled with little or no growth in membership during the past several years, Environmental Action staffers constantly faced the financial pressures linked to communicating with their members and, in the end, succumbed to them.

Environmental Defense Fund

The Environmental Defense Fund remains committed to the newsletter format, although the length of the *EDF Letter* has doubled from four to eight pages in recent years. In 1991–1992 EDF members received five newsletters, down from seven in 1977–1978.[50] These newsletters continue to be heavily laden with policy issues, among them many of the salient environmental issues of the day (e.g., global warming, clean air, wetlands, deforestation [international], ozone depletion, ocean dumping, Everglades, and alternative fuels). Overall, the number of issues presented dropped from fifty to thirty-one from the 1977–1978 analysis to the present time frame. As was the case in the earlier analysis of EDF, it is less than clear how this type of communication influences members. Nonetheless, the pattern of truncation of messages is evident in EDF communications as well.

Leaders, Members, and Attentive Citizens: Hearing the Echo of Your Own Voice?

Significant changes have taken place during the last two decades regarding the communications between leaders of environmental organization and members and also in their communications to the broader attentive public. Without membership data comparable to those presented earlier for 1978, it is difficult to reach any formal conclusions about the consequences of the changes in frequency and quality of the communications. Nevertheless, several implications are plausible, based on the relationships found between communications from the leadership and the membership's opinions and activities in the earlier time period.

First, the efforts of the National Wildlife Federation to provide more policy information may pay long-term dividends at the expense of short-term costs. Informing one's membership has its costs. With one of the broadest policy agendas in the environmental movement, NWF is involved in most of the important environmental and conservation issues of the day. Up until recently this fact was not widely known among NWF members. Even today leaders inform their members in general terms for fear of losing members with specific policy differences. As a result of the expanded policy agenda (at least by NWF standards), members are better informed and may act upon such information. However, with the expansion of the agenda delivered to members, the Federation has also suffered a significant drop in membership in recent years. Certainly there is no single direct relationship that explains the decline in membership. As demonstrated in chapter 2, many environmental organizations felt the impact of the economic downturn in the early 1990s, and NWF was no exception. Be that as it may, National Wildlife editor Robert Strohm has concluded from his recent editorial experiment that he may have taken the magazine a bit too far in the policy direction (even at the 12 percent level) and intends on scaling back the number of policy articles in future issues.[51]

Aside from the changes in *National Wildlife,* each of the remaining four communications: *Sierra, Wilderness, Environmental Action,* and *EDF Letter* has undergone a significant truncation in the frequency of distribution and in the number of issue messages sent to members. A negative conclusion regarding the agenda-setting process by leaders for members seems inevitable. Now, it may well be the case that these communications are no longer meant to serve such a purpose and that broader, more public, strategies are employed.

The Sierra Club, for example, may be employing such a strategy in marketing *Sierra* magazine. In order to market a magazine to an audience beyond the Sierra Club membership base, organization leaders are repackaging the product to match consumer desires, but not without significant internal tensions. Since their magazine was made available to the general public, the level of conflict has escalated between policy staffers, seeking to inform Sierra Club *members* of ongoing policy initiatives, and the editors of *Sierra*, seeking to gain new *subscribers* to the magazine.[52] The resulting diluted message provides fewer policy cues to active members who, in the past, have demonstrated their willingness and ability to act on their own accord for political purposes, often in response to the direct and indirect cues received in communications.

The most recent attempts by organization leaders to inform and mobilize members and nonaffiliated citizen activists have involved the creation of websites on the internet.[53] In addition to individual websites, there are scores of information directories that lead internet users to relevant information sources. There are more than eighty national and international environmental organizations with websites on the worldwide web, including the four existing organizations analyzed here, and a number of environmental directories through which the websites may be located (see table 6.9).

In 1998 there are 100 million people linked to the internet, with tens of thousands of new subscribers being added every month.[54] Environmental organizations are not alone in their quest for wider dissemination of political information. In the United States, the National Rifle Association, for example, has had great success using the internet. According to Glen Caroline, director of Grassroots Programs for the NRA Institute for Legislative Action, the lobbying arm of NRA, its website, http://www.nra.org, transmits 25,000 files per day to interested members but also to interested citizens who do not belong to the NRA. In addition, the NRA has developed an e-mail mailing list that contains 23,000 NRA subscribers; these active members may be contacted on a daily basis with information and instructions regarding grassroots mobilization. Through its website and e-mail network, the NRA delivers approximately three million files of information to active citizens each month.[55]

While the website as a means of communication is a passive one from an organizational perspective in that users or members must initiate the communication, it is clearly a cost-effective method of providing information to a larger audience attentive to a particular cause or issue. Each of the four organizational websites identified above provides a wealth of information about the organization's structure, finances, issue agenda, policy activ-

TABLE 6.9 Environmental Websites and Directories
Websites

ORGANIZATION	ADDRESS*
Environmental Defense Fund	http://www.edf.org
National Wildlife Federation	http://www.nwf.org
Sierra Club	http://www.sierraclub.org
The Wilderness Society	http://www.wilderness.org

Directories

Econet
Envirolink
Environmental Organization WebDirectory
GreenNet
Galaxy/Community/Environment
WWW Virtual Library: Environment
Yahoo Society and Culture: Environment and Nature

*http: (HyperText Transport Protocol)—the commonly accepted protocol for
transporting hypertext documents on the worldwide web.

ities, and additional group-specific materials. In most cases, the organizations' newsletters and magazines are also available to be downloaded to local printers attached to the internet.

From a democratic perspective, the worldwide web has provided an important means through which active citizens may gather the information they seek to participate more effectively in our political system. From an organizational perspective, however, there is a potentially serious problem with the broad dissemination of information, as people not paying dues have access to the same information (e.g., magazines, newsletters, and other selective incentives or premiums) as do members of the organizations. For these environmental organizations, there is no differentiated access to their websites. Furthermore, revenues generated by selling access to the internet (through Prodigy, America on Line, Netscape, etc.) are not shared by individual website owners.

For example, as a member of the Environmental Defense Fund, I receive five or six issues of *EDF Letter* and several additional solicitation letters and telephone calls for my annual membership fee. By logging on to my computer and gaining access to the worldwide web, I am able to visit the EDF website without any additional cost. Once I am in the website, I have access not only to current and back issues of *EDF Letter* but also to a wide variety of interesting and useful information, including annual reports, staff biographies, and up-to-date policy information generated by EDF researchers and lobbyists. There is policy information in the website that I do not receive as a dues-paying member of the organization. If I belong to EDF solely to gather information that is useful to become better informed on environmental matters, it is not clear why I should continue to write checks to EDF when I have access to all of the information I need or want, free of charge through the internet. This is the twenty-first-century version of Olson's free-rider problem.

From my discussions with the leaders of these organizations as well as with a host of other public interest leaders whose organizations have websites with unrestricted access, it is clear that most of these individuals have not contemplated this potential problem. Certainly, if the underlying goal of organizations pursuing a collective good is broad political representation in the political arena, then the outreach programs through the internet are positive steps. However, membership development leaders may be well advised to keep a weather eye on the impact of these internet connections on members and nonmembers in the years to come. In addition to the potential free-rider problem, there remains the inherent weakness in the use of the internet for political mobilization purposes mentioned earlier. As a means of mobilization and activation, the use of websites relegates leadership to a role of passive participant, unable to signal to members proactively if and when rapid, direct action is needed.

In an era when constituency mobilization is perhaps the most important means of influencing the policy-making process, it is ironic that the organizations that transformed grassroots lobbying into one of the vital and necessary elements of influencing policy makers over the last three decades, have, for financial and editorial reasons, moved away from the primary means of linking leaders and members for political purposes—namely, direct communications. It will become increasingly difficult for organized interest groups to mobilize grassroots activists if they continue to diminish the role of communicating policy from leadership to membership in a timely fashion. As a result, it will become much harder for policy makers to hear the *echoes* in Washington if members cannot hear the *voices* of their leaders.

7

Organizational Leadership and Grassroots Empowerment
Reinvigorating Public Interest Representation in the United States

For it so falls out,
That what we have we prize not to the worth
Whiles we enjoy it, but being lack'd and lost,
Why, then we rack the value, then we find
The virtue that possession would not show us,
Whiles it was ours.
—WILLIAM SHAKESPEARE, *Much Ado About Nothing,* ACT 4, scene 1

The public interest sector in the United States has indeed become a major political enterprise, with more than four billion dollars raised and spent annually on maintaining thousands of organizations and providing political representation for millions of citizens. Today, the public interest sector is dramatically different from the loosely organized social movements of three decades ago. As is evident from the analyses in the preceding chapters, the organizational transformations that have occurred have raised many questions regarding the representative nature of these transformed political enterprises, in particular about the linkages between organizational voices and membership echoes. Public interest organizations are attempting to formulate strategies that allow them to maintain their organizational infrastructures as well as to provide some form of public interest representation. In years past, the measure of public interest representation was the presence or absence of an organizational process and structure that linked leaders with their supporters. As the public interest sector has evolved over the last three decades, this measure is no longer a valid indicator of political representation in the contemporary political environment. Organizational structures and processes may provide some indication of the priorities established by the organizations; however, if organizational outputs are incorporated into the analysis of political representation, organizational attributes other than democraticizing mechanisms in the policy-making process are increasingly important.

In each of the preceding chapters various manifestations of the links between organizational maintenance and political representation are presented. In chapter 1, the organizational growth and professionalization across the entire public interest sector were analyzed. In the late 1990s public interest organizations are representing a broad range of societal interests. As these organizations matured, they have become less reliant on support from private foundations, the chief catalysts for organizational growth three decades earlier. Memberships that grew dramatically in the 1980s and leveled off in the early 1990s are again showing signs of remobilization; the percentage of organizations with memberships greater than 200,000 almost tripled in less than fifteen years. With the growth in financial support and membership has come professionalized staffing. Recent additions to organization staffs are much more likely to be people professionally trained for the positions they fill in the organizational technostructure.

However, with all of the growth in the public interest sector, individual organizations remain constrained by government regulation. New and revised laws on lobby registration and taxes limit the amount of time and money organizations may spend on direct policy advocacy—i.e., on lobbying. This particular organizational constraint is one of the more difficult problems to resolve; nonetheless it reaches to the representational core of public interest organizations. These groups often attract members by stressing the eminent threats arising from government or industry decisions or "nondecisions." Yet, 501(c)(3) organizations, which comprise roughly two-thirds of the public interest sector, may spend only 20 percent (or less) of their organizational resources on congressional lobbying. There are a growing number of organizations that have managed to find some resolution to the lobbying constraint problem by reorganizing to become "full-service" public interest organizations. This may be the organizational structure for the next century, allowing public interest contributors and members to receive "more bang for the buck" in direct policy advocacy.

The consequences of professionalization in the leadership ranks of the public interest sector are also analyzed. As the leaders of these organizations are increasingly drawn from the professional ranks in government, business, or the sector itself, the bias against women grows proportionally. Women hold only 20 percent of the executive directorships in the public interest sector. As the value of previous high-level government experience rises, the prospect for women serving in those leadership positions diminishes, as the gender bias in one sector of society is transferred to another. Representation on the boards of directors of public interest organizations also reflects the sectorwide bias. Women serving on these governing boards account for less than one-quarter of the total board membership of public interest organizations.

In terms of making the governing public interest bureaucracy more representative by including more women, minorities, and younger people, the organizations find themselves in continual searches for such "representative" candidates for service. Some of the larger organizations have resorted to hiring talent search agencies to track down qualified women and minorities. Regardless of the tactics employed, public interest organizations are seeking to open up the governance process to new leaders with new ideas by changing the mechanisms through which boards are chosen. Professional competence as well as management skill and financial input are necessary characteristics for board members. Elected boards tend to be used as reward systems for time well served in the organization. Given the transformation that has taken place in the public interest sector, there is little room for any type of spoils system in the professional management of public interest organizations.

Nevertheless, there still exists a small social network from which the governing elites are currently being drawn. In interview after interview, board members and organization leaders argued that the public interest sector is too "cliquish." The old boy network remains alive and well on the governing boards of many public interest organizations. Regardless of the sources, the public interest governing elite must be broadened. The same small circle of friends is not sufficient to lead the public interest sector into the next century.

In addition to bringing in new, more diverse individuals to the governing ranks, the problem of board (and staff) stagnation is a problem to be dealt with in order to reinvigorate the leadership ranks. Lifetime service on governing boards is not healthy for the long-term well-being of public interest organizations. Some mechanism that provides for circulation of governing elites will serve the broader interest of these organizations. Similarly, it is necessary to have leadership circulation. For example, when William Turnage accepted the position of executive director at The Wilderness Society in 1978, he told the board that he would stay with the organization for no more than seven to eight years—he stayed for seven. This is a healthy approach to public interest leadership. This is not to say that stability is not welcome. On the contrary, The Wilderness Society had hired and fired three executive directors in the four years before hiring Turnage and has had a similar pattern of turnover following his departure. There is little benefit in this type of leadership turnover. However, more stable transitions in power are a positive indicator of organizational well-being.

Related to the problem of organizational representativeness is the issue of organizational governance. Chapter 3 details the variety of organizational forms and leadership management styles within the national environmental movement. Each of the five environmental organizations has a distinctive approach to dealing with the balance between organizational maintenance and political representation. The Sierra Club presents the most grassroots-oriented approach to organizational governance. With its elected boards, direct policy input from membership through the organizational infrastructure, and its newly streamlined consensus-building decision-making process, the Sierra Club is one of the more democratized organizations in the public interest sector. However, the consensus-building process, while perhaps more democratic, may not be any more representative in the larger policy-making arena than other organizational approaches.

The Sierra Club persists due, in large part, to its infrastructure. However, the levels of membership participation in voting for board members, attendance at chapter and group meetings, and participation in the Club's massive outings and travel programs are 10 to 20 percent or less. Therefore, many of the organization's members resemble in motivations and actions the members of other public interest organizations, such as Public Citizen, the American Civil Liberties Union, or Mothers Against Drunk Driving. They are solicited by mail and respond by writing their membership checks once a year. The only difference between these contributors is that Sierra Club members receive an attractive magazine as a premium for their membership check.

Up until recently, the National Wildlife Federation has made little effort to link its members to the internal policy-making process; rather, it seeks to "educate" citizens on the environment in the most general terms. While the Federation may be organized like a labor union, its policy-making process involves state leaders only in an indirect manner. The Federation is a prime example of an environmental conglomerate. With more than 600 staff professionals, the organization has one of the broadest issue agendas in the national environmental movement. Yet, the survey findings presented in chapters 5 and 6 indicate that the Federation members are the least informed on relevant policy issues of any of the environmental organization memberships. Perhaps the new leadership of Mark Van Putten will reverse this trend.

Far from an environmental conglomerate, Environmental Action met with its collective demise in October of 1996. The organization never attracted more than 30,000 members. Run as a collective, the leadership corps was small and underfunded. Regarding the relationship with the membership, the leaders of Environmental Action did their best to keep their members informed on all of the policy issues targeted by the staff. Unfortunately the lack of funding precluded any systematic and timely efforts at mobilizing membership support in the final years. In addition, its change in tax status, from 501(c)(4) to 501(c)(3), further limited its ability to act as a strong advocate in the policy and electoral arenas.

There are a number of other groups like Environmental Action that are floundering; they barely subsist. Perhaps the merging of such organizations may result in more effective policy advocacy for the movement as a whole. By sharing the heavy organization costs while retaining the environmental fervor of these smaller organizations, new, more viable organizations may provide greater diversity for the environmental movement in the late 1990s, rather than further redundancy.

The Environmental Defense Fund has experienced the most growth in recent years, with current membership numbering more than 300,000. Organizationally, it has responded in a number of ways in order to facilitate this rapid growth. Organization leaders have diversified the fund-raising operations to include not only contributions from individual members—by offering premiums previously underutilized by the organization—but also funds from a wide variety of foundations and other institutions. Its working philosophy of market-based environmentalism has clearly attracted a wide audience of support, even if the organization is controversial among the national environmental leaders.

Finally, The Wilderness Society provides an excellent case study of organizational transformation and its possible consequences. Following several years of internal turmoil, The Wilderness Society board selected William Turnage as its new leader in 1978. Upon his arrival, he discarded the existing organizational structure and employee base and replaced it with a professionalized, hierarchical leadership structure and staff. His goal was not simply to maintain the organization, but to provide effective policy advocacy in the political forum. By stressing leadership, management, and professionalization, he served his constituents more effectively than had any of the strategies employed by the earlier Wilderness Society leadership.

After Turnage left The Wilderness Society, organizational leadership passed to George Frampton, a more transactional leader, able to continue on the course set forth by Turnage. Upon Frampton's departure, Karin Sheldon directed the organization until Jon Roush was identified as the new leader in 1994. Roush, much like Turnage, was a political entrepreneur with a number of new approaches to organizational leadership. Unlike the circumstances under which Turnage took control, however, the organization was not in dire financial straits when Roush arrived. Consequently, the transformational leadership style of Roush was met with much greater organizational resistance. After all, Roush could not and did not fire the one hundred employees he inherited, as Turnage had done when he arrived sixteen years earlier.

By 1996 Roush's personal problems as well as his relationships with his staff in Washington and across the country were adversely affecting his ability to lead The Wilderness Society effectively. As a result he left the organization. Today, The Wilderness Society, under the direction of William Meadows, continues to search for ways to revitalize the organization in the policy debate for the next century.

Effective organizational leadership is a vital component in the maintenance and representation of public interest organizations. Without leadership, such organizations are destined to languish as secondary actors in the political arena and, more important, serve to drain the larger movement of scarce resources by attracting members and their contributions.

Effective leadership entails not only the coherent articulation of a policy agenda to government officials and to members of the organization but also the ability to attract and retain a sustainable membership base. Direct-mail marketing, telemarketing, and, to a lesser extent, canvassing, are the methods of choice in the public interest sector for dealing with the constant process of recruitment and retention of members. These methods of mobilization have led to the growing problem of organizational membership overlap, particularly in each of the public interest movements, but also in the sector more generally. As a result, membership loyalties are often short-lived. When these consumers of public interest commodities are inundated with literally hundreds of pieces of mail a year from a wide variety of organizations, they cannot help but engage in comparison shopping.

For example, assume that last year an individual entered the public interest sector by writing checks to five public interest organizations. In the last twelve months, he or she might easily have received direct-mail solicitations from an additional thirty organizations. This public interest consumer also receives at least five renewal notices from the five organizations joined last year.[1] The options available to this individual are numerous—renew all five and add several more; drop some or all memberships and add one or more new ones; drop some or all memberships and add no new organizations; or exit the sector completely by renewing none of the memberships. Based on retention studies conducted by direct-marketing specialists and the organizations themselves, the most likely option is the renewal of membership in *some* of the initial organizations and adding one or two new organizations. Recall the comment made by list broker Richard Hammond: "The white liberal community in this country is always looking for change." Members of public interest organizations are a distinctive breed. Throughout the environmental movement, recognition of this distinctiveness has led to serious rethinking of the environmental agenda and has raised broader issues, such as "environmental racism," to prominence on the policy agenda of some organizations.[2] Organizationally, leaders seek to understand this distinctiveness in order to serve their constituents better. For the Sierra Club, for example, this attentiveness is demonstrated in the overhauling of *Sierra* magazine. Anybody may now purchase *Sierra* in bookstores and at newspaper and magazine outlets. The new format makes the magazine attractive to a wider audience. As a result, however, the Sierra Club is now faced with the subscriber-member dilemma. Clearly the editorial staff wants to sell magazines, but the organization leaders want members.

The issue is not yet resolved, but the very fact that this type of problem even arises highlights the difficulty of maintaining a *stable* base of support, whether in the form of members, contributors, or subscribers. There are several ways in which the public interest sector could strengthen its position in terms of membership maintenance. First, the most difficult problem organizations face is retaining first-year members. The most successful organizations have only 60 to 70 percent retention rates for new members; the least successful retain less than 40 percent of first-year members. After the second year of membership, the retention curve rises dramatically. Only recently have some organizations begun offering initial two-year memberships at a reduced rate. Since it has been demonstrated in all organizations that second-year retention rates are ten to fifteen percentage points higher than first-year retention rates, the two-year membership approach, while more costly in the short run, may pay greater dividends in the long-term retention of members.

The techniques of direct mail and telemarketing must also be complemented by additional methods of recruitment. Returning to some of the social movement approaches would not hinder the progress of the public interest sector for the next century. Some organizations are attempting to motivate existing members to reach out to their friends to bring more people into these organizations. Outreach is particularly necessary among the younger people in this country, given the graying problem identified in chapter 4. To most of the membership and development personnel interviewed, younger people (college students) are difficult to maintain. First, they tend to want something for their money—activities, premiums, outings. Second, they have little money and therefore must be offered student rates. Third, by nature they are a transient sector of society—they are difficult to track. In general they represent a short-term hassle for most organizations. While significant efforts in recent years have begun the mobilization of a new generation of environmentalists, the opinions of some organization leaders remain rather short-sighted on this issue.

The public interest sector will wither away if significant cultivation of the younger generations does not continue. And given the expected cycle of activism in the 1990s noted in the comments by Ralph Nader in chapter 1, now is not the time to ignore the younger cohort predisposed toward revitalizing the public interest sector. Consequently, other more direct means of mobilization are necessary. Few methods are as cost effective as direct mail, but the long-term viability of public interest advocacy calls for significant investment in the future. A growing number of environmental organizations have recognized this fact in recent years and are implementing campus outreach programs as well as helping to shape environmental curricula at all educational levels.[3]

In chapter 5, the motivations of members of environmental organization and the incentive structures provided by their organization leaders are analyzed. Since many individual members in the environmental samples were recruited through direct mail, it is necessary to evaluate the degree to which this method of recruitment and retention affects the representative nature of the relationship between leaders and members.

Until very recently most of the public interest organizations using direct mail as the primary means of membership recruitment were doing very little systematic analyses of their memberships. While some organizations have included "member surveys" in their recruiting materials, these surveys are not valid and reliable measures of membership opinions, due to the selection biases among the respondents and the very low response rates. Through the 1980s, organizations were spending millions of dollars on direct-mail recruiting, yet little or no money on attempting to understand the motivations and opinions of their members. Certainly the individual direct-mail packages sent to prospective members were test-marketed by the organizations or their contracted direct-mail specialists; however, few groups surveyed their memberships outside the context of recruitment and retention.

There are some important exceptions to this pattern. The American Association of Retired Persons for decades has conducted numerous national surveys each year on a wide range of policy issues affecting citizens aged fifty and older. Even with this attentiveness to their constituency, AARP leaders have at times failed to interpret their own survey results accurately. The passage and later repeal of the Medicare Catastrophic Coverage Act of 1988 provides an excellent example of organizational leadership misreading the intensity of concern of a vocal minority of their membership.[4]

Their own internal polling showed that wealthier senior citizens were not supportive of the legislative language that would have charged them additional premiums for catastrophic coverage based on their income.[5] The AARP leadership, arguing that the vast majority of the members would benefit from the legislation, wholeheartedly supported the passage of the act. Following its passage, AARP leaders suffered the wrath of their wealthy members. They had clearly underestimated the magnitude of displeasure expressed by this vocal minority. As a result of the backlash, more vocal senior citizens groups returned to Congress in 1989 and lobbied for the repeal of the legislation that had been championed by the AARP only months earlier.

Like the AARP, a growing number of public interest organizations are becoming increasingly attentive to the opinions of their members, through membership survey research as well as focus groups. For example, one of the top Democratic polling firms in Washington, Lauer, Lalley, Victoria, Inc., conducts membership surveys and focus groups for a variety of public interest organizations, including the Nature Conservancy, The Wilderness Society, Earth Share, the National Parks and Conservation Association, and Families USA.[6]

In addition to the trend toward increased awareness of membership preferences, organizations increasingly use high-technology direct-mail research firms that are able to target potential members in ways other than simply renting the mailing lists of allied organizations. Based on the knowledge gained by organizations through their own internal polling, development directors are able to construct demographic profiles of their membership, much like the analysis presented in chapter 5.

Today, organization leaders are able to present their profiles to this new breed of direct-mail firm, and the firm will produce names, addresses, and telephone numbers of people fitting the profiles. By purchasing consumer information, credit card data, voting records, motor vehicle registration data, marriage and divorce records, and virtually any information made public by government or made available for purchase in the private sector, the direct-mail firms are able to target with great accuracy individuals with distinctive political, social, economic, and demographic characteristics. Thomas Halatyn of Campaign Mail & Data, Inc., states that his firm has more than 100 million individual voter files and more than 160 million individual consumer files, with many files including fifty or more variables.[7]

With this resource available, development directors are able to request from these firms very distinctive lists of potential donors. For example, an organization could request a list of 10,000 names with the following characteristics: white, female, married with grown children out of the home, Gold Card holder, home owner, registered Democrat, household income over $100,000, purchases merchandise through catalog mail-ordering, contributes money to political campaigns, wears glasses, reads *Newsweek,* and drives a foreign car.

For a premium price, these firms will produce such a list for the organization. The disadvantage for the organization is that the list will cost much more than the usual list rental rate of $75 per 1,000 names quoted in chapter 4. The advantage is that the response rate will be much higher than the 2- or 3-percent level of the rental list, perhaps as high as 15 to 20 percent. Using such firms raises the maintenance costs of direct-mail recruitment, at least in the short term, but it has the potential to eliminate some of the waste associated with the bulk mailing of millions of direct-mail pieces in order to achieve a 2 percent response rate.

In the context of membership recruitment, an analogy may be drawn to the political campaign industry. It is often said that 75 (80, 90) percent of all campaign expenditures are wasted—they simply have no political impact. The trick is to identify the effective 25 percent. This continues to be the task of political consultants and campaign managers. The same problem is faced by organizations that use a variety of organizational incentives and recruiting tools to maintain memberships. Much of what organizations provide as incentives or mail to prospective members is wasted. To resolve these and other issues related to recruitment and retention, organization leaders are becoming more informed about their existing membership bases. As development directors and membership recruiters slowly learn from their earlier mistakes, the task of organizational maintenance through direct mail and other complementary methods is becoming a more fruitful investment.

Finally, chapter 6 addressed the complexity of the ongoing relationships between leaders and members. The question faced by organization leaders is simple: to inform or not to inform. From a representative perspective, the obvious resolution is to tell members as much as possible about the policy agenda of the organization and to place that agenda in the political context of the current policy-making environment. Even if the communication process exists only from leaders to members, members are at least equipped to respond to the actions of their leadership. However, the majority of public interest organizations, by virtue of their 501(c)(3) tax status, are legally constrained in their efforts to mobilize their members. The specific policy content that directs members to act is considered to be part of the organizational lobbying effort; as a result, the research, writing, and editorial time is counted against the 20 percent limit imposed on 501(c)(3) organizations. This is yet another reason to adopt a "full-service" organizational structure.

The introduction of the worldwide web into the public interest communications process raises some important issues for organization leaders and their links not only with their members but with the larger attentive public in the United States as well. From a democratic perspective, the more informed the citizenry, the stronger the ties between those who govern and the governed. From an organizational perspective, however, the presentation of group information to a wider audience reduces the value of that information as a selective incentive for members and, therefore, may have consequences for long-term organizational maintenance.

More controlled two-way communications on the internet, such as through e-mail, may be the more effective route to informing activist members. The information disseminated by organization leaders is targeted to specific members, with responses communicated in a similar fashion. Regardless of the strategy employed, the worldwide web cannot be overlooked in the organization and representation of political interests in the next century.

It is important for leaders of public interest organizations to realize that except for the members of the larger education-oriented organizations such as National Wildlife Federation, people belonging to public interest organizations are a major part of the attentive public sector in this country. Membership in these organizations is only one manifestation of their political activism. In order to capitalize on this activism, through grassroots lobbying strategies, effective organizational communications provide a clear indication of the policy agenda pursued by the leadership. To be sure, organizations run the risk of losing members due to disagreements on policy issues. However, from a policy perspective, the benefits of an informed membership outweigh the costs.

Challenges for the Twenty-First Century

Clearly, national environmental organizations have had more demonstrable success in the area of coalition and alliance building during the past decade than they have had in reestablishing a vigorous and active grass-roots lobbying force. As explained in chapters 3 and 4, the leadership of national environmental organizations has joined forces to promote sound environmental policies or to impede the efforts of the conservative wing of the Republican leadership in Congress to dismantle environmental, health, and safety regulations. These leaders have also developed significant relationships with corporate and philanthropic financial interests. In addition, some national organizations have also pursued governmental entities for funding support through contracts and grants. The environmental movement is not unique in this growing coalitional character. According to an analysis of alliance behavior across the spectrum of private sector, government, and nonprofit sector entities conducted by Booz-Allen and Hamilton, four times as many alliances were formed in the United States between 1987 and 1992 as were formed between 1980 and 1987.[8]

Marketing researchers George Milne, Easwar Iyer, and Sara Gooding-Williams more recently looked specifically at the coalitional activities of environmental organizations.[9] Using a national organizational directory published by the National Wildlife Federation,[10] the research team analyzed a sample of 195 environmental organizations. The mean annual budget for the sample was $9 million and the organizations averaged 63,400 members. The research sought to identify "alliances formed by environmental organizations with 1) government agencies, 2) for-profit businesses, and 3) other environmental organizations."

Overall, they identified two-thirds of the sample organizations as having ongoing alliances with other environmental groups, just over half (53 percent) having alliances with government or regulatory agencies, and 30 percent having formal relationships with for-profit businesses.[11]

These alliances reflect the new pragmatism of the late 1990s wherein the "us" and the "them" are not as clearly defined anymore as they were in the past. Beyond alliances or coalitions with business, government, and non-profit organizations, the national environmental movement has also felt the impact of the growing corporate and philanthropic sector. While the influence of the foundation world at the outset of the new environmental era in the late 1960s was documented in chapter 2, the rise of environmental philanthropy in the 1990s, coinciding with the downturn in grassroots membership support, has had a significant impact on national organizations.

Many of the new environmental philanthropists come to these newly forged relationships with agendas that may or may not complement existing organizational policy strategies. According to environmental reporter

Scott Allen, "the rise of environmental philanthropists is creating a new—and controversial—class of powers behind the throne who are shaping the movement with their money. . . . This growing stream of 'big money' support has buoyed the environmental movement even as grassroots contributions have faltered, but in the eyes of some it raises a troubling question: Are the funders now calling the shots?"[12]

In 1997, 15 percent of the Sierra Club budget was derived from "major donors." In the same year, the Nature Conservancy received at least $1 million each from General Motors, Dow Chemical, and Enron Corp., an energy company, following on the heels of a $5 million donation from Paul Allen, cofounder of Microsoft, in 1996. The charitable trusts and foundations have caught the environmental bug in recent funding cycles and giving to environmental causes increased 91 percent from the mid-1980s to the mid-1990s, whereas giving to all causes increased by only 67 percent. In all, more than $400 million annually reach the treasuries of American environmental causes via large charitable donations, including $22.5 million from the Pew Charitable Trusts, $17 million from the Ford Foundation, $16.7 million from the John D. and Catherine T. MacArthur Foundation, $9.8 million from the W. Alton Jones Foundation, and $8.2 million from the Heinz Endowments in recent annual giving. Teresa Heinz, manager of the endowments left by her late husband, Senator John Heinz, also donated $447,923 to the Environmental Defense Fund in 1993, where she serves as vice-chairperson on the board.[13] Support from these trusts and foundations has even spawned environmental organizations without members, such as the Environmental Working Group, a Washington-based environmental research and media outreach entity, funded by the Tides Foundation and other institutional contributors.

The impacts of these alliances with environmental philanthropists, together with the increasing ties to corporate interests through direct gifts and donations, merchandising contracts, advertising arrangements, and product endorsements, as well as links to governmental entities through contracts and grants are obviously varied but nonetheless potentially troubling. Alliances and coalitions among environmental organizations at the national level have proven successful in both the proactive movement of environmental policy initiatives in years past and in blocking perceived antienvironmental legislation in recent years. These other alliances and relationships, while often financially beneficial to the organizations, underscore the reason for the broader perception that the national environmental movement is "too corporate, too Washington." These dollar-driven dealings have led Jeff St. Clair, editor of the *Wild Forest Review,* to conclude: "the environmental movement is now accurately described as just another cynical, well-financed special interest group."[14]

While there is an ultimate economic bottom line in nonprofit management of national environmental organizations, there are potentially negative consequences for organizational leaders who get caught up in the money chase, one of which is abandonment by the grassroots membership. Organizational leaders must balance the need for financial resources in large increments with the need to present an organizational image that is worth the investment of tens or hundreds of thousands of annual membership fees by individuals. These national organizations will cease to exist without the support of individual members.

As active as many of these organizations have been in cultivating alliances and coalitions, they have been equally inactive at cultivating their grassroots memberships. While steps have been taken in the last year or two in several of the national organizations to rectify the lack of communication between national leadership and the members, there is still a long way to go. The recent internal debate over national immigration policy within the ranks of the Sierra Club, for example, may well serve as a catalyst for increased communications between leadership and membership, although there is still tension between the grassroots and the Washington leadership in particular.[15] Gina Neff, writing in *The Nation*, summarizes the feelings of some Sierra Club members: "The Sierra Club leadership, not wanting to sacrifice its cozy relationship with Clintonian environmentalism, would never have opposed NAFTA or supported the ban on commercial logging in national forests had it not been forced to do so by its insurgent ranks."[16] Again, the cry is "too Washington." While these charges may be a bit harsh, there is a kernel of truth here that may be applied equally to virtually all of the national environmental organizations attempting to voice collective policy preferences in the national policy-making arena. From what the preceding two chapters have shown about members of environmental organizations, it is clear that these citizen activists seek to be empowered by a variety of political resources, including the public interest organizations they support as advocates. Members of public interest groups are drawn from the sector of society that has the willingness and ability to think about and act on issues with long-term consequences. These are individuals who understand the meaning of the opening quotation—Shakespeare's late-sixteenth-century version of Joni Mitchell's 1960s elegy: "You don't know what you've got 'till it's gone." In many ways, they do know what they've got and wish to protect it before it's gone. James Q. Wilson, quoted at the beginning of chapter 5, echoes this point: "The prospect of a loss is more likely to motivate action than the expectation of a gain."[17] As a result, members of public interest organizations are "extrarational" in an economic sense.[18] They are motivated by purposive goals (and threats) to pursue collective goods. As such, they seek to be informed about the policies they are supporting.

For organization leaders, the words of Shakespeare and Mitchell may be interpreted a bit differently. In a significant manner, membership comes to mind when applying these words to organization leaders. That is, leaders did not know what they had (members), or did not appreciate their importance in the rapid growth years of the 1980s, until they lost them in the early 1990s. The mission for organization leaders in the late 1990s and beyond is the empowerment of their memberships to become more effective actors in the policy-making process. Grassroots education and outreach programs are increasingly important means through which the agendas of public interest activists are acknowledged and acted upon by policy makers in Washington. Given the relative lack of salience of environmental issues in the 1996 presidential and congressional elections and once again in the 1998 congressional elections, it is doubly important to maintain a constant and vigilant presence in the policy-making process, both on the part of the grassroots and of the organization representatives in Washington.

Outside the national environmental movement, there is an excellent example of organization leadership responding to the need to reinvigorate its representative capacity so as to maintain an effective presence in Washington and, more important, in congressional districts across the United States. Common Cause, under the leadership of Ann McBride, has sought to reorganize itself in order to become a more effective grassroots advocacy organization. Prior to its internal restructuring, the organization maintained paid professional directors in all fifty states, regardless of whether the state had 1,500 members or 15,000 members.

The Common Cause leadership and governing board ruled that maintaining this structure was not the most effective way to allocate resources. As a result, grassroots resources are now distributed to states and districts based on membership representation. Today, some states no longer have paid directors, but other states and localities with high concentrations of members are better served by the organization. The decision to reallocate organizational resources was not easily made. Nonetheless, it reflects the political realities of public interest representation in the late 1990s.[19] There are more than a handful of nationally organized public interest organizations that could benefit from such an internal restructuring.

Despite efforts to expand the bases of support for public interest causes in the United States, public interest representation remains inherently elitist in nature. The leaders are drawn from the social, economic, and political elite strata of society. Members of public interest organizations themselves are demographically representative of the middle or upper classes. Nonetheless, the representational relationship between organization leaders and members need not be "virtual" in a Burkean sense. While the realities of the political influence game make "acting for" the interests of members a necessity, public interest representation is more effective when organization leaders are "acting with" their members, perhaps not in a direct manner, but in essence. Leadership implies followership; this is what distinguishes powerful individuals from leaders. If the public interest sector is directed by leaders, then the essence of political representation as "acting with" is achieved. If members are informed and act upon that information, then representation as "acting with" is strengthened.

Citizens may also play a significant role in resolving the issue of meaningful political representation. Public interest representation should not be a spectator sport on the part of an organization's members. Organization members must "guard the guardians" as well as act upon the policy agendas of organization leaders. Not all public interest organizations are effectively managed, professionally organized, and politically effective. In fact, many are not. My advice for members of public interest organizations (or potential members) is *caveat donator*, let the donor beware. Public interest organizations in the late 1990s continue to wrestle with the relationships of organizational maintenance and political representation. There are numerous organizations that are so entrenched in the struggle to maintain an organizational existence that political representation is no more than a dream. These are the public interest money pits. Watchdog organizations, such as the American Institute of Philanthropy, provide an important public service as it is quite difficult to distinguish the money pit from the effective organization simply by analyzing the direct mail sent by public interest organizations.

Given the magnitude of the issues raised in the public interest sector as well as the costs associated with advocating policy issues in Washington today, it is vital to preserve the representational aspect of public interest advocacy. Simple organizational existence advances no cause. Public interest organizations for the late 1990s and beyond must refocus their attention on the policy-making process by linking citizen activists as advocates for organizational goals in a meaningful manner. They must make policy advocacy their primary goal and structure their organizations in order to magnify the sounds of public interest *voices and echoes*.

Public Interest Organizations
Foundation for Public Affairs

Methodological Note on Organization Selection

Earlier versions of tables 1.1–1.4 were published in "More Bang for the Buck: The New Era of Full-Service Public Interest Organizations," in Allan J. Cigler and Burdett A. Loomis, eds. *Interest Group Politics*, 3d edition. The explanation of the comparability of the data sets provided in that chapter, modified and reproduced below, is equally applicable to the analyses presented here: The Berry data presented in chapter 1 were drawn from chapters 2 and 3 of *Lobbying for the People*. The Foundation for Public Affairs data sets were drawn from the foundation's publications, *Public Interest Profiles* (1986–1987, 1992–1993, 1996–1997). For each public interest group that is presented below, seventy-eight variables were coded from the raw data presented in each profile.

To analyze further the comparability of the two data sources, those organizations included in both sets (N = 35) were analyzed and data for an additional fourteen organizations not found in the Foundation for Public Affairs publication were pieced together. The tables presented in this chapter were replicated using data from the forty-nine groups available for both data sources as well as from additional data gathering. Aside from the finding in table 1.1 on year of origin, the patterns uncovered in the full analysis of all Berry groups and FPA groups are comparable to those in the smaller "panel" study. Any variation in the two data sources reflects changes made in the public interest sector during the intervening thirteen years between the two studies. Sample sizes reported for the tables in chapter 1 differ due to missing values in the data sets.

FOUNDATION FOR PUBLIC AFFAIRS, 1986–1987

Business/Economic

Center on Budget and Policy Priorities
Center for the Study of American Business
Citizens for a Sound Economy
Citizens for Tax Justice
Competitive Enterprise Institute
National Bureau of Economic Research
National Center for Employee Ownership
National Center on Occupational Readjustment, Inc.
National Tax Limitation Committee
National Taxpayers Union
Tax Foundation, Inc.
Tax Reform Research Group
Work in America Institute

Civil/Constitutional Rights

American-Arab Anti-Discrimination Committee
American Association of Retired Persons
American Civil Liberties Union
American Life League
Amnesty International USA
Children's Defense Fund
Citizens Committee for the Right to Keep and Bear Arms
Gray Panthers Project Fund
Handgun Control, Inc.
Leadership Conference on Civil Rights
League of United Latin American Citizens
Mexican American Legal Defense and Education Fund
Migrant Legal Action Program
NAACP Legal Defense and Educational Fund, Inc.
National Abortion Rights Action League
National Association for the Advancement of Colored People
National Association of Arab Americans
National Council on the Aging, Inc.
National Council of La Raza
National Council of Senior Citizens
National Organization for Women
National Rifle Association of America

National Right to Life Committee, Inc.
National Right to Work Committee
National Urban League
Nine to Five, National Association of Working Women
Operation PUSH, Inc.
People for the American Way
Planned Parenthood Federation of America, Inc.
Public Service Research Council
Southern Christian Leadership Conference
Women's Equity Action League
Women's Legal Defense Fund

Community/Grassroots

American Agriculture Movement
American Coalition for Traditional Values
Association of Community Organizations for Reform Now
Campaign for Economic Democracy
Center for Community Change
Citizen Action
Citizen's Choice
Eagle Forum
Illinois Public Action Council
Institute for Local Self-Reliance
Midwest Academy/Citizen's Leadership Foundation
Mothers Against Drunk Driving
National People's Action/National Training and Information Center
National Trust for Historic Preservation
New York Public Interest Research Group
Ohio Public Interest Campaign
Partners for Livable Places
People for the Ethical Treatment of Animals
Rural America
Students Against Driving Drunk
United States Public Interest Research Group
Veterans of Foreign Wars of the United States

Consumer/Health

Action on Smoking and Health
American Council on Science and Health
Buyers Up

Center for Auto Safety
Center for Science in the Public Interest
Center for Study of Responsive Law
Citizen's Utility Board, Inc.
Community Nutrition Institute
Consumer Alert
Consumer Federation of America
Consumers Union of United States, Inc.
Food Research and Action Center Inc.
Group Against Smokers' Pollution
Health Research Group
International Organization of Consumers Unions
National Consumers League
National Insurance Consumer Organization
Public Voice for Food and Health Policy

Corporate Accountability/Responsibility

Business Committee for the Arts, Inc.
Catalyst
Center for Business Ethics
Center for Corporate Public Involvement
Center for Public Resources
Committee for Economic Development
Council on Economic Priorities
Council on Foundations Inc.
Council of Institutional Investors
Ethics Resource Center, Inc.
The Foundation Center
Independent Sector
INFACT
Institute for Educational Affairs
National Alliance of Business
National Committee for Responsive Philanthropy
Partnerships Data Net, Inc.
VOLUNTEER—The National Center

International Affairs

American Committee on Africa
American Israel Public Affairs Committee
American Security Council

The Arms Control Association
Atlantic Council of the United States
Business Executives for National Security
Caribbean/Central American Action
Center for Defense Information
Center for National Security Studies
Center for Strategic and International Studies
Coalition for a New Foreign and Military Policy
Committee for National Security
Council on Foreign Relations
Council for a Livable World
Foreign Policy Association
Global Tomorrow Coalition
Institute for International Economics
Physicians for Social Responsibility
SANE (Committee for a Sane Nuclear Policy)
TransAfrica
Trilateral Commission
Union of Concerned Scientists, Inc.
Washington Office on Latin America

Media

Accuracy in Media
Action for Children's Television, Inc.
Center for Investigative Reporting
Foundation for American Communications
The Media Institute
National Coalition on Television Violence
Reporters Committee for Freedom of the Press
Scientists' Institute for Public Information
Telecommunications Research and Action Center

Political/Governmental Process

Advocacy Institute
American Conservative Union
American Legislative Exchange Council
Americans for Democratic Action
Business-Industry Political Action Committee
Center for Responsive Politics
Citizens Against Government Waste

Citizens Against PACs
Citizens' Research Foundation
Committee on Political Education
Common Cause
Congress Watch
The Conservative Caucus, Inc.
The Council of State Governments
Democrats for the 80's
Fund for a Conservative Majority
League of Women Voters of the U.S.
Liberty Federation
National Center for Initiative Review Foundation
National Committee for an Effective Congress
National Conference of State Legislatures
National Congressional Club
National Conservative Political Action Committee
National Governors' Association
National League of Cities
National Women's Political Caucus
Project on Military Procurement
Project VOTE!
Public Citizen, Inc.
Responsible Government for America Foundation
U.S. Conference of Mayors
Young Americans for Freedom

Public Interest Law

Alliance for Justice
Capital Legal Foundation
Center for Law in the Public Interest
Center for Law and Social Policy
Gulf and Great Plains Legal Foundation
HALT, Inc.
Mid-America Legal Foundation
Mid-Atlantic Legal Foundation, Inc.
Mountain States Legal Foundation
National Legal Center for the Public Interest
Pacific Legal Foundation
Public Advocates, Inc.
Public Citizen Litigation Group

Southeastern Legal Foundation, Inc.
Trial Lawyers for Public Justice
Washington Legal Foundation

Religious

American Friends Service Committee
American Jewish Congress
Anti-Defamation League of B'nai B'rith
Clergy and Laity Concerned
Institute on Religion and Democracy
Interfaith Center on Corporate Responsibility
National Association of Evangelicals
National Conference of Catholic Bishops/United States Catholic
    Conference
National Council of the Churches of Christ in the U.S.A.

Environmental

Citizen/Labor Energy Coalition
Citizens for a Better Environment
Clean Sites, Inc.
Clean Water Action Project
The Conservation Foundation
Critical Mass Energy Project
Defenders of Wildlife
Environmental Action, Inc.
Environmental Defense Fund, Inc.
Environmental Law Institute
Environmental Policy Institute
Friends of the Earth
Greenpeace U.S.A.
Izaak Walton League of America
League of Conservation Voters
National Audubon Society
National Campaign Against Toxic Hazards
National Coalition Against the Misuse of Pesticides
National Wildlife Federation
Natural Resources Defense Council, Inc.
The Nature Conservancy
Resources for the Future
Sierra Club

Solar Lobby
U.S. Committee for Energy Awareness
The Wilderness Society
World Resources Institute
World Wildlife Fund

FOUNDATION FOR PUBLIC AFFAIRS, 1992–1993

Business/Economic

Center on Budget and Policy Priorities
Center for the Study of American Business
Citizens for a Sound Economy
Citizens for Tax Justice
Committee for a Responsible Federal Budget
Competitive Enterprise Institute
Council on Competitiveness
National Taxpayers Union
Work in America Institute

Civil/Constitutional Rights

AIDS Coalition to Unleash Power (ACT UP)—New York
American-Arab Anti-Discrimination Committee
American Association of Retired Persons
American Civil Liberties Union
American Life League
Amnesty International USA
Center for Democratic Renewal and Education, Inc.
Center for Policy Alternatives
Children's Defense Fund
Citizens Committee for the Right to Keep and Bear Arms
The Fund for the Feminist Majority
Handgun Control, Inc.
Human Rights Campaign Fund
Lambda Legal Defense and Education Fund, Inc.
Leadership Conference on Civil Rights
League of United Latin American Citizens
Mexican American Legal Defense and Education Fund
NAACP Legal Defense and Educational Fund, Inc.
National Abortion Rights Action League
National Association for the Advancement of Colored People

National Association of Arab Americans
National Coalition for the Homeless
National Council on the Aging, Inc.
National Council of La Raza
National Council of Senior Citizens
National Organization for Women
National Rifle Association of America
National Right to Life Committee, Inc.
National Urban League
Nine to Five, National Association of Working Women
People for the American Way
Planned Parenthood Federation of America, Inc.
Pro-Life Action League
Women's Legal Defense Fund

Community/Grassroots

American Family Association
The American Society for the Prevention of Cruelty to Animals
Association of Community Organizations for Reform Now
Center for Community Change
Citizen Action Fund
Community for Creative Non-Violence
Concerned Women for America
Eagle Forum
Family Research Council
Institute for Local Self-Reliance
Midwest Academy
Mothers Against Drunk Driving
National Center for Neighborhood Enterprise
National People's Action/National Training and Information Center
Nuclear Free America
Parents' Music Resource Center
U.S. ENGLISH Foundation, Inc.
United States Public Interest Research Group

Consumer/Health

Action on Smoking and Health
American Council on Science and Health
American Foundation for AIDS Research (AmFAR)
Bankcard Holders of America

Buyers Up
Center for Auto Safety
Center for Biomedical Ethics
Center for Science in the Public Interest
Center for Study of Responsive Law
Community Nutrition Institute
Consumer Alert
Consumer Federation of America
Consumers Union of United States, Inc.
Co-op America
Food Research and Action Center Inc.
Health Research Group
International Organization of Consumers Unions
Motor Voters
National Anti-Vivisection Society
National Charities Information Burea
National Consumers League
National Insurance Consumer Organization
National SAFE KIDS Campaign
National Safe Workplace Institute
Public Voice for Food and Health Policy

Corporate Accountability/Responsibility

The Business Enterprise Trust
Business-Higher Education Forum
Catalyst
Center for Business Ethics
The Center for Corporate Community Relations at Boston College
Center for Corporate Public Involvement
Center for Public Resources
Committee for Economic Development
Council on Economic Priorities
Council on Foundations
Council of Institutional Investors
Ethics Resource Center
The Foundation Center
Independent Sector
INFACT
National Alliance of Business
National Committee for Responsive Philanthropy

The National VOLUNTEER Center
United Shareholders Association

International Affairs

American Committee on Africa
American Israel Public Affairs Committee
The Atlantic Council of the United States
Center for Defense Information
Center for Strategic and International Studies
Council for a Livable World
DataCenter
Federation for American Immigration Reform (FAIR)
Global Tomorrow Coalition
Human Rights Watch
Institute for International Economics
Physicians for Social Responsibility
SANE/FREEZE: Campaign for Global Security
TransAfrica
Union of Concerned Scientists, Inc.
The Washington Institute for Near East Policy

Media

Accuracy in Media
Action for Children's Television, Inc.
Fairness and Accuracy in Reporting
The Media Institute
Media Research Center

Political/Governmental Process

Advocacy Institute
The American Conservative Union
American Legislative Exchange Council
Americans for Democratic Action
Arab American Institute
Business-Industry Political Action Committee
Center for Responsive Politics
Citizens Against PACs
Citizens' Research Foundation
Common Cause
Congress Watch

Congressional Management Foundation
The Conservative Caucus, Inc.
Council for Excellence in Government
The Council of State Governments
Fund for a Conservative Majority
League of Women Voters of the U.S.
National Association of Attorneys General
National Committee for an Effective Congress
National Conference of State Legislatures
National Governors' Association
National League of Cities
National Women's Political Caucus
OMB Watch
Public Citizen, Inc.
The U.S. Conference of Mayors
Vote America Foundation
Women's Campaign Fund

Public Interest Law

Alliance for Justice
Capital Legal Foundation
Christic Institute
HALT—An Organization of Americans for Legal Reform
Landmark Legal Foundation
The Mid-America Legal Foundation
Mountain States Legal Foundation
National Legal Center for Public Interest
Pacific Legal Foundation
Public Citizen Litigation Group
Southeastern Legal Foundation
Trial Lawyers for Public Justice
Washington Legal Foundation

Religious

American Jewish Committee
American Jewish Congress
Anti-Defamation League of B'nai B'rith
Interfaith Center on Corporate Responsibility
National Association of Evangelicals

National Conference of Catholic Bishops/United States Catholic
  Conference
National Council of the Churches of Christ in the U.S.A.

Environmental

Center for Marine Conservation
Citizens Clearinghouse for Hazardous Wastes, Inc.
Clean Water Action
Coalition for Environmentally Responsible Economies
Conservation International Foundation
Critical Mass Energy Project
Defenders of Wildlife
Earth First!
Earth Island Institute
Environmental Action, Inc./Environmental Action Foundation
Environmental Defense Fund, Inc.
Environmental Law Institute
Farm Animal Reform Movement (FARM)
Foundation for Research on Economics and the Environment
Foundation on Economic Trends
Friends of the Earth
The Fund for Animals
Greenpeace U.S.A.
The Humane Society of the United States
In Defense of Animals
INFORM, Inc.
Izaak Walton League of America
League of Conservation Voters
The Keystone Center
League of Conservation Voters
National Audubon Society
National Coalition Against the Misuse of Pesticides
National Parks and Conservation Association
National Toxics Campaign
National Wildlife Federation
Natural Resources Defense Council
The Nature Conservancy
People for the Ethical Treatment of Animals
Pesticide Action Network (PAN) North America Regional Center

Rainforest Action Network
Renew America
Resources for the Future
Rocky Mountain Institute
Sea Shepherd Conservation Society
Sierra Club
Student Environmental Action Coalition (SEAC)
The Wilderness Society
Wildlife Enhancement Council
World Resources Institute
World Wildlife Fund

FOUNDATION FOR PUBLIC AFFAIRS, 1996–1997

Business/Economic

Americans for Tax Reform
Center for the New West
Center for the Study of American Business
Center on Budget and Policy Priorities
Citizens for a Sound Economy
Citizens for Tax Justice
Committee for a Responsible Federal Budget
Competitive Enterprise Institute
The Conference Board, Inc.
Council on Competitiveness
Family and Work Institute
National Taxpayers Union

Civil/Constitutional Rights

American-Arab Anti-Discrimination Committee
American Civil Liberties Union
Americans United for Separation of Church and State
Asian American Legal Defense and Education Fund
Center for Democratic Renewal
Children's Defense Fund
Handgun Control, Inc.
Human Rights Campaign
Lambda Legal Defense and Education Fund, Inc.
Leadership Conference on Civil Rights
League of United Latin American Citizens

Mexican American Legal Defense and Education Fund
NAACP Legal Defense and Educational Fund, Inc.
National Abortion and Reproductive Rights Action League
National Association for the Advancement of Colored People
National Association of Arab Americans
National Coalition for the Homeless
National Congress of American Indians
National Council of La Raza
National Council of Senior Citizens
National Council on the Aging, Inc.
National Organization for Women
National Organization on Disability
National Rainbow Coalition
National Rifle Association of America
National Urban League
Nine to Five, National Association of Working Women
Operation Rescue
People for the American Way
Planned Parenthood Federation of America, Inc.
Women's Legal Defense Fund

Community/Grassroots

American Family Association
The American Society for the Prevention of Cruelty to Animals
Association of Community Organizations for Reform Now
Center for Community Change
Citizen Action Fund
Concerned Women for America
Eagle Forum
Family Research Council
Mothers Against Drunk Driving
National Center for Neighborhood Enterprise
National Coalition Against Domestic Violence
National Training and Information Center/National People's Action
U.S. ENGLISH Foundation, Inc.
United States Public Interest Research Group

Consumer/Health

Action on Smoking and Health
Advocates for Highway and Auto Safety

AIDS Action Council
American Association of Retired Persons
American Council on Science and Health
American Foundation for AIDS Research (AmFAR)
Center for Auto Safety
Center for Biomedical Ethics
Center for Science in the Public Interest
Center for Study of Responsive Law
Coalition on Smoking OR Health
Consumer Alert
Consumer Federation of America
Consumers International
Consumers Union of United States, Inc.
Co-op America
Families USA Foundation
Food Research and Action Center
National Charities Information Burea
National Consumers League
Public Voice for Food and Health Policy

Corporate Accountability/Responsibility

Business for Social Responsibility
Catalyst
Center for Business Ethics
Coalition for Environmentally Responsible Economies
Committee for Economic Development
Council of Institutional Investors
Council on Economic Priorities
Council on Foundations
Ethics Resource Center
The Foundation Center
Independent Sector
INFACT
National Alliance of Business
National Committee for Responsive Philanthropy
The Points of Light Foundation

International Affairs

American Israel Public Affairs Committee
Amnesty International USA

Center for Defense Information
Center for Strategic and International Studies
Council on Foreign Relations
DataCenter
Federation for American Immigration Reform (FAIR)
Human Rights Watch
Institute for International Economics
Peace Action
Physicians for Social Responsibility
TransAfrica
Union of Concerned Scientists, Inc.
The Washington Institute for Near East Policy
Zero Population Growth

Media

Accuracy in Media
Center for Media and Public Affairs
Center for Media Education
Fairness and Accuracy in Reporting
Media Access Project
Media Research Center
National Coalition on Television Violence

Political/Governmental Process

Advocacy Institute
The American Conservative Union
American Legislative Exchange Council
Americans Back In Charge Foundation/Term Limits Legal Institute
Americans for Democratic Action
Arab American Institute
Business-Industry Political Action Committee
Center for Public Integrity
Center for Responsive Politics
Citizens' Research Foundation
Common Cause
Concord Coalition
Congressional Management Foundation
Council for Excellence in Government
Council of State Governments
Empower America

League of Women Voters of the United States
National Conference of State Legislatures
National Governors' Association
National League of Cities
National Women's Political Caucus
Project Vote Smart/Center for National Independence in Politics
Public Citizen, Inc.
U.S. Term Limits
U.S. Conference of Mayors
United We Stand America
Women's Campaign Fund

Public Interest Law

Alliance for Justice
Atlantic Legal Foundation, Inc.
Institute for Justice
Landmark Legal Foundation
Mountain States Legal Foundation
National Legal Center for Public Interest
Pacific Legal Foundation
Southern Poverty Law Center
Trial Lawyers for Public Justice
Washington Legal Foundation

Religious

American Jewish Committee
American Jewish Congress
Anti-Defamation League of B'nai B'rith
Christian Coalition
Focus on the Family
Interfaith Center on Corporate Responsibility
National Conference of Catholic Bishops/United States Catholic
    Conference
National Council of the Churches of Christ in the U.S.A.
Traditional Values Coalition

Environmental

American Rivers
Center for Marine Conservation
Citizens Clearinghouse for Hazardous Wastes, Inc.

Conservation Fund
Conservation International
Defenders of Wildlife
Earth First! Journal
Earth Island Institute
Environmental Defense Fund, Inc.
Environmental Law Institute
Environmental Working Group
Farm Animal Reform Movement (FARM)
Foundation on Economic Trends
Friends of the Earth
The Fund for Animals
Greenpeace U.S.A.
Humane Society of the United States
In Defense of Animals
Izaak Walton League of America
The Keystone Center
League of Conservation Voters
National Audubon Society
National Coalition Against the Misuse of Pesticides
National Parks and Conservation Association
National Trust for Historic Preservation
National Wildlife Federation
Natural Resources Defense Council
The Nature Conservancy
People for the Ethical Treatment of Animals
Pesticide Action Network (PAN) North America Regional Center
Rainforest Action Network
Resources for the Future
Rocky Mountain Institute
Sea Shepherd Conservation Society
Sierra Club
The Wilderness Society
World Resources Institute
World Wildlife Fund

Operationalization of Variables (chapter 5)

Premiums:

(V42)   I enjoy (name of group's magazine or newsletter) and/or other benefits of membership like outings very much. (VI I N U
VU: coded 5 4 3 2 1)

(V47)   I personally gain much from the information I receive from
(group). (VI I N U VU: coded 5 4 3 2 1)

(V62)   The (name of group's magazine) magazine is the major reason
why I belong to (group). (SA A N D SD: coded 5 4 3 2 1)

Goals:

(V45)   My contribution to (group) is helping to influence government action on conservation/environmental problems. (VI I
N U VU: coded 5 4 3 2 1)

(V49)   Without contributions like mine, (group) would be unable to
work for improved environmental quality. (VI I N U VU:
coded 5 4 3 2 1)

(V50)   If (group) achieves its goals, my life and children's lives will
directly benefit. (VI I N U VU: coded 5 4 3 2 1)

Interactions:

(V46)   I belong to (group) because of the encouragement of my
friends. (VI I N U VU: coded 5 4 3 2 1)

(V48)   Many knowledgeable and influential people support (group).
(VI I N U VU: coded 5 4 3 2 1)

Threats:

(V$_{43}$)   Some important aspects of my life are threatened by environmental/conservation problems. (VI I N U VU: coded 5 4 3 2 1)

(V$_{44}$)   Solving environmental problems is so important that I try to support any effort aimed at that goal. (VI I N U VU: coded 5 4 3 2 1)

(V$_{53}$)   If we don't act now to preserve the environment, things will get much worse. (VI I N U VU: coded 5 4 3 2 1)

Environ/   How appropriate or inappropriate would the following de-
Conserv: scriptions be if applied to you personally? "Environmentalist" (VA A N I VI: coded 5 4 3 2 1) "Conservationist" (VA A N I VI: coded 5 4 3 2 1)

Income: Total family income before taxes (coded 1–10)

Education: What is the highest grade you have finished and gotten credit for in regular school, college, or graduate school? (coded 1–20)

Religion: Catholic, Protestant, Jewish coded 1; No organized religion, but ties with nature and none coded 0

Party: Strong Democrat-Strong Republican (coded 7–1)

Ideology: Strong Conservative-Middle of Road-Radical (coded 1–6)

Mail: Received a membership appeal in the mail coded 1; through friends, received membership as a gift, saw magazine and sent for membership coded 0.

Efficacy:

(V$_{276}$)   I don't think public officials care much what people like me think. (SA A N D SD: coded 1–5)

(V278)   Sometimes politics and government seem so complicated that a person like me can't really understand what's going on. (SA A N D SD: coded 1–5)

(V281)   People like me don't have any say in what the government does. (SA A N D SD: coded 1–5)

Occup: Dummy variable; Professional, technical and kindred workers, managers and administrators (except farm) sales workers coded 1; other occupations coded 0.

Numgrps: Number of groups the respondent reports belonging to (from an inclusive list of 31 groups)

Activity: How often do you do the following:

(V242)   Work with others to help solve community problems

(V245)   Contact public officials or politicians

(V247)   Attend political rallies or meetings

(V248)   Spend time working for a political party or candidate

Serious: Overall, how serious are the nation's environment problems? (We are rapidly approaching disaster-very serious-serious-somewhat serious-not very serious-problems basically solved: coded 6 5 4 3 2 1)

Sex: 1 = Male 0 = Female

Age: In what year were you born? (transformed into actual age).

Validity and Reliability (chapter 5)

The interpretations of the results presented assume that the measures incorporated in the analyses are both valid and reliable. The models tested include four theoretically distinct dimensions (variables): PREMIUMS, GOALS, INTERACTIONS, and THREATS. Concern for the validity and reliability of these concepts is particularly important, given the motivational character of the variables and the fact that these theoretical concepts have not yet been analyzed empirically in this form.

Validity refers to the extent to which the measures of the variables correspond to the concepts they are intended to reflect. Generally, there are four basic approaches to validation. First, pragmatic validation, assessing how well the measure works in allowing one to predict behaviors and events, requires alternative indicators of the concepts that are themselves valid indicators. Here no such measures are available; unfortunately, there are few instances in social science research where valid alternative indicators exist. The second type of validation, construct validation, is more applicable to this analysis. There are two types of construct validation: internal or convergent and external.[1] Convergent validation involves devising several measures of the same variable and assessing the degree of correlation among the similar measures. Multiple indicators are employed for each concept. For the PREMIUMS, GOALS, and THREATS variables, three indicators were used. For the INTERACTIONS variable, only two indicators were available from the survey data. The correlation matrices for each concept are presented in table C.1. The average correlations are: PREMIUMS indicators (.415), GOALS indicators (.452), THREATS indicators (.396), and INTERACTIONS indicators (bivariate) (.494). All correlations are significant at .0001 level. These correlations suggest that the indicators used in the construction of each of the conceptual variables are

TABLE C.I Multiple Indicators: Bivariate Correlations

CONVERGENT VALIDATION

	Premiums				Goals		
	V_{42}	V_{47}	V_{62}		V_{45}	V_{49}	V_{50}
V_{42}	1.000			V_{45}	1.000		
V_{47}	0.554	1.000		V_{49}	0.497	1.000	
V_{62}	0.380	0.321	1.000	V_{50}	0.442	0.418	1.000
	Interactions				Threats		
	V_{46}	V_{48}			V_{43}	V_{44}	V_{53}
V_{46}	1.000			V_{43}	1.000		
V_{48}	0.494	1.000		V_{44}	0.396	1.000	
				V_{53}	0.348	0.443	1.000

focused on the same dimension. The lack of additional related variables with valid measures precludes external validation. Obviously future research efforts must incorporate such valid measures into their designs.

The third approach to validation is referred to as discriminant validation. This type is dependent upon the degree to which a measure of a given concept allows one to distinguish that particular concept from other distinct concepts. In this analysis, it is necessary to assess the degree to which the theoretical concepts as operationalized are valid measures of distinct motivational dimensions. Table C.2 provides the correlation matrix for the four concepts. The pattern that emerges provides evidence of distinctiveness among concepts, but also of the interrelationships between them. It appears that there are two sets of two related dimensions. The INTERACTIONS and PREMIUMS dimensions are correlated to a degree as are GOALS and THREATS dimensions to a greater degree. Theoretically, one would expect to find relationships between these dimensions. For example, one of the indicators used for the PREMIUMS measure includes outings as a benefit. While these activities may be tangible selective incentives, the theoretical argument presented identifies an affective component in outings and other group-related activities. Beyond this, the INTERACTIONS dimension is underspecified. No measure of the symbolic connection between individual and group is offered in the sample responses.

Even though there may be little in the way of symbolic attraction or sense of "oneness" brought about by group symbols, slogans, or leaders within the environmental sector, a more complete operationalization of this dimension should include such an indicator. Andrew McFarland, in his study of Common Cause, presents data from a 1974 poll of members that shows 13 percent of the sample joining the organization because of its leader, John Gardner.[2] One would expect to find a similar pattern among the membership of Ralph Nader's Public Citizen. As for organization symbols, one struggles to relate specific symbols to specific environmental organizations. The Nature Conservancy's oak leaf design seen across the nation on countless Volvo station wagons may be one of the few recognizable symbols in the environmental movement. The environmental movement, more generally, had its Earth Day symbol that emblazoned the movement's flag in the late 1960s and early 1970s, as did the anti-Vietnam Peace movement, with its "peace sign." These symbols faded as the movements became a series of distinct organized (public) interests.[3]

The distinction between the GOALS and THREATS dimensions is less clear in the analysis, as one would expect it to be, given that the normative dimension is argued theoretically to be more of a socially imposed precondition that motivates certain citizens to take group-related action. The GOALS dimension directs active individuals to particular organizations. Even with a sample of actual organization members, however, the correlation is not overwhelming—suggesting that there are indeed two distinct dimensions at work. The argument for their distinctiveness lies in the source of the incentive. GOALS incentives are provided by individual groups; whereas THREATS incentives are provided by the individual's larger social context. The distinction between these two dimensions is presented in this analysis not so much for the additional explanatory power of both but for theoretical inclusiveness. Future studies incorporating both members and nonmembers may find the THREATS dimension to be a rather strong predictor in the process of distinguishing potential organization adherents (as well as actual members) from the free riders, which Olson identifies as the much larger of the two groupings.

Finally, theoretical concepts may be validated at face value. The indica-

TABLE C.2 **Discriminant Validation**

	PREMIUMS	GOALS	INTERACTIONS	THREATS
Premiums	1.000			
Goals	−0.071[**]	1.000		
Interactions	0.354[***]	0.042[*]	1.000	
Threats	−0.015	0.509[***]	0.052[*]	1.000

tors used in the operationalization of the theoretical concepts are fairly self-evident and are supported by the theoretical literature cited in the construction of the motivational dimensions. In summary, there is evidence supporting the valid operationalization, through multiple indicators, of individual concepts and somewhat weaker evidence in support of the distinctiveness of the concepts. Given the unique nature of this study, there is sufficient evidence to accept the validity of the concepts presented. As for the reliability of the concepts and their measures, the validity of the concepts leads one to consider them reliable as well, for a measure cannot be valid without being reliable. Reliability is jeopardized by random error. One may test the reliability of the measures by what is known as subsample method. The data collection method used here is an excellent example of the subsample method. Five distinct subsamples were surveyed using the same (with group name changes) survey instrument. By comparing the correlation matrices for the four dimensions for each group, one is able to test the stability of the measures. In virtually all cases, the bivariate correlations between the dimensions were not significantly different from the correlations presented in table C.2.

Specific and General Policy Issues Presented in 1977–1978 Communications

Aside from the amount of space given to policy concerns as well as the number of items presented, it is important to evaluate the actual issues raised in order to assess the breadth of the environmental agendas set by the group leaders. Obviously, the sheer magnitude of the policy content in these interest group communications says much about the significance given the agenda-setting process. But by looking at the general and specific issues presented, one can more easily assess the degree to which an agenda is set and the breadth of issues upon which leaders requested membership activity.

In 1977–1978 the National Wildlife Federation possessed a comparatively weak environmental agenda, at least in the communications process with its members. Salient issues included: the Alaska Lands bill (HR39, S1500), Water Projects (all five organizations mention these two issues), the Carter Energy Package, Repeal of the 1872 Mining Laws, Redwoods Park Expansion, strip-mining, and the Tellico Dam (cf., media coverage presented in chapter 2). No issue seems to dominate the agenda even though the organization's actual issue agenda is quite broad. Moreover, general issues include a variety of policy concerns, with no clear direction.

In many cases, the specific policy issues were presented retrospectively; issues were raised after legislative action had occurred. Final outcomes were reported, leaving members no chance to react to the issues. Concerning items about which members were asked to contact congressmen or executive officials, there were never more than one request for such action on a specific issue (although on the water projects issue members were asked to write to members of Congress and to the president). It is difficult to discern any pattern of salience among the issues. One must remember, however, that in the case of NWF, there is no intent to use the magazine for activist mobilization. With such a broad membership, NWF leaders cannot afford to use *National Wildlife* as a mobilizing tool.

The late John Strohm, founding editor of *National Wildlife* and its editor until his recent death, stressed the educational aspect of the magazine but recognized that the magazine was more than that. "It gives the Federation a showpiece, an identity, a public relations tool and it also makes them money without advertising."[4] NWF board member John Gottschalk is a bit more blunt on the topic: "The magazine is pap. I think the magazine does a great job, but it certainly does not very often get into any of those very controversial issues which could have a tendency to polarize the membership. They try to stay with bland things that everybody loves."[5] Even Strohm himself was willing to grant that the magazine had a certain softness to it: "You must never, never, never underestimate members' intelligence, but you must never overestimate their comprehension." It is clear that the National Wildlife Federation views the communication process between leaders and members quite differently from the vast majority of public interest organizations.

The Sierra Club, on the other hand, had a different perspective on group communications and hence a much more extensive agenda, judging from the number of issues, as well as some degree of hierarchy among the issues, given the multiple presentations of issues. In general, the salient issues were those involving concerns for wilderness maintenance, broadly defined. Looking at the specific issues raised, SC members were presented with several articles and notes on each of several wilderness issues, including: Outer Continental Shelf, Endangered American Wilderness Act, Expansion of the Redwoods National Park, Clean Air Act (including Class I air quality status for National Parks), the Alaska Lands bill, and Mineral King (Sequoia National Park). SC also provided information on numerous general policy issues with some attention focused on wilderness issues.

However, like NWF, Sierra Club also faced the dilemma of addressing a large (if more highly educated and affluent) audience (now more than one-half million members). According to former *Sierra* editor Jim Keough:

> At least 40 percent [of Sierra Club members] never do a thing; they never do anything but send us money for their membership, that's their participation in the Club. What we're trying to do is to at least get them to read the magazine which would be then a second bit of participation in the Club. To do that we can't focus in on Club priorities and argue them in the magazine because unless you are really hooked into them, they're boring as hell. So our goal is to try to pick out representative examples of

the priorities. . . . Maybe it's just trying the old *dolce mutilare* approach. We want people to be entertained and interested with the magazine; we want them also to be informed.[6]

While the Sierra Club seems generally focused on wilderness issues, The Wilderness Society issue agenda is thoroughly embedded in wilderness concerns. The dominant issue component of the wilderness agenda in 1977–1978 was the passage of the Alaska Lands bill (HR30, S1500). Between the two communications, *Living Wilderness* and *Wilderness Report,* the Alaska bill was presented to the TWS membership no less than twenty-five times. Even more significant are the twenty-five instances where members were requested to contact members of Congress and even Secretary of the Interior Andrus. No other group blanketed the issue as did TWS.

What made the organization's efforts at activating members so effective was that the leaders continuously tracked the bill through the legislative process. Members were asked to write members of Congress on numerous occasions: when the bill was introduced in the House, when it was introduced in the Senate, when it reached the committees for debate, when administration officials testified at committee hearings, when it came to the House and Senate floors for a vote, and (after the period of the content analysis) when it reached the president's desk. On a dozen occasions, members were asked to voice their support for a strong Alaska Lands bill. TWS demonstrated on this bill, on the Endangered American Wilderness Act (HR3454), and on the Montana Wilderness Study Bill (HR393) how interest group communications may serve as efficient activating mechanisms and agenda setters. TWS has a dominant priority in its lobbying agenda—wilderness preservation—and it makes that priority known to its membership.

As stated earlier, TWS has dropped its publication *Wilderness Report* due to financial problems. However, problems with the Internal Revenue Service have also limited the amount of policy content in more recent issues of *Wilderness* magazine. As a 501(c)(3) tax-exempt non-profit organization, The Wilderness Society is limited in the amount of "lobbying" activity it may undertake. The IRS considers the presentation of specific policy issues in publications to be a part of the lobbying process. Hence, in 1985 the IRS began to pay close attention to the policy content of the organization's publications. Patricia Byrnes, managing editor of *Wilderness,* explained the lengths to which the IRS has gone in order to ascertain whether or not TWS is meeting tax guidelines. "We have to fill out time sheets on what we have done, so if I had something here [in the magazine] that is lobbying, and the IRS sees other people in the Conservation Department working on the same issues, they are going to add it up and say 'Wait a minute!'"[7] Of the five organizations analyzed, only the Sierra Club remains unburdened by the IRS regulations for 501(c)(3) organizations regarding grassroots lobbying efforts.

Unlike TWS, Environmental Action had a much broader issue agenda, including several very salient issues (at least for this group). If anything, EA leaders may have overwhelmed their members with environmental/political issues. Theirs was a much more diverse issue agenda than that of the other groups. Covering issues from transportation, noise pollution, solid waste management, utility rates, toxic chemicals, and waterway user fees to the more conventional environmental issues of wilderness protection, strip-mining, water and air pollution, alternative energy sources (solar, geothermal), public lands, and endangered species, EA did not miss any of the "environmental action" going on in Washington in 1977–1978. Indeed, they were involved, and involved their members in other political matters such as the Panama Canal debate, the MX and B-1 controversies, and legislative action to control recombinant DNA research as well as one of the group's major environmental/political issues, nuclear proliferation.

Aside from the specific policy issues addressed, EA did much more for their members in terms of facilitating their efforts to contact public officials. For example, the entire April 3, 1977, issue, entitled "How to Write Congress . . . in One Easy Lesson," is directed at grassroots lobbying efforts. Included in this and other issues during the year is a section called "Where to Write." In each section, members are provided with the names and addresses of other environmental lobbying groups, executive agencies, and relevant committees and subcommittees in the House and Senate. With their very attentive membership, EA also was comparatively successful in some of its efforts, given the more extreme positions taken by group leaders and the relatively small size of the group membership.

EA is perhaps financially the weakest of the five organizations, which causes problems in the communications process. Today, instead of twenty-four issues a year, EA offers only four issues annually. What they have lost in this cutback is the ability to inform their members about relevant policy issues as they are being discussed in Congress. EA editors are constrained in their efforts to link leaders and members in their agenda-setting process. Rose Audette, former editor of *Environmental Action*, understood the difficulty of representing members when the ability to do so in a meaningful way did not exist. "If you're an honest group, which I think Environmental

Action is, we try to represent our members, but it's real hard, because if you don't have the money you can't find out what your members want." However, even if the opportunity to set an issue agenda in this manner did exist, she is not sure it would be the most effective way. "You could certainly send out questionnaires and get back answers and tabulate them and say that is democracy. But is that how an organization should set its agenda in a politically charged world like this? . . . I'm not sure, I'm not sure at all. It's an interesting question—how do you do that?"[8]

While there are significant differences in the magazines of NWF and EA, Audette's comments reflect at least some ambivalence regarding the internal agenda-setting process that Joel Thomas commented on earlier. However, as a mechanism for promoting further external political participation of group members, EA clearly makes the best use of its communications. If one wishes to learn about what EA is doing, one simply needs to read *Environmental Action*. The same cannot be said of NWF and its magazine.

Finally, the Environmental Defense Fund does not present an issue agenda to its members. The legislative agenda is particularly sparse with only seven issues mentioned, only three of which came before Congress. The executive branch issues are quite focused and are not directly relevant to the salient legislative environmental issues under debate in Congress at the time. Similarly, the general issues presented are not a part of any cohesive environmental agenda.

Today, EDF still faces a communications dilemma. The format and frequency of its newsletter remains largely unchanged. Nonetheless, EDF has made significant efforts to increase its visibility. A recent innovation in the organization's marketing strategy is the production of a direct video program, "Environmental Vision: Twenty Years of Innovative Solutions," that is being sent to members in order that they may proselytize their friends and neighbors. By all accounts EDF has been quite successful, as its membership has increased sixfold in the past decade, despite a comparatively truncated communications process.

It is evident that the use of interest group communications as cuing mechanisms and/or agenda setters is not consistent across the five organizations. EDF and NWF do a comparatively poor job of setting a coherent agenda, but agenda setting for members may not be the purpose of their communications. The TWS and SC agendas are generally focused on wilderness issues. Finally, Environmental Action presents an extensive and diverse issue agenda.

Operationalization of Variables (chapter 6)

Runaway Technology, (V231): We have already let technology run away from us. 2 (Strongly Agree) 1 (Agree) −1 (Disagree) −2 (Strongly Disagree)

Wilderness Preservation, (V225): Further additions to the nation's protected wilderness should have a high priority for the federal government. 2 (SA) 1 (A) −1 (D) −2 (SD)

Alternative Technology, (V229): The key to our problems is to develop "alternative" or "soft" technologies which are non-polluting and low energy and resource consuming. 2 (SA) 1 (A) −1 (D) −2 (SD)

Population Problems, (V191): What are the most important cause of environmental problems? Please rate each of the following potential causes as to their importance . . . Population growth. 2 (VI) 1 (I) −1 (U) −2 (VU)

Opposition to Nuclear Power, (V200): Taking into account all you have heard or read, how do you feel toward nuclear power plants in general? −2 (Very Favorable) −1 (Fairly Favorable) 1 (Fairly Unfavorable) 2 (Very Unfavorable)

Animal Protection, (V215): An endangered species must be protected, even at the expense of commercial activity. 2 (SA) 1 (A) −1 (D) −2 (SD)

Air Pollution, (V185): How serious do you think air pollution is in this state? 2 (Very) 1 (Fairly) −1 (Not So) −2 (Not Serious)

Water Pollution, (V186): How serious do you think water pollution is in this state? 2 (Very) 1 (Fairly) −1 (Not So) −2 (Not Serious)

Energy Conservation, (V199): How serious would you say the energy situation is in the U. S.? 2 (Very) 1 (Fairly) −1 (Somewhat) −2 (Not Serious)

Toxic Substances, (V209): Recently the government banned certain food dyes used in things like cherries and hot dogs because of studies showing that they might cause cancer. Do you generally approve or disapprove of what the government did? 2 (Strongly Approve) 1 (Approve) −1 (Disapprove) −2 (Strongly Disapprove)

Preface

1. V. O. Key Jr., *The Responsible Electorate: Rationality and Presidential Voting, 1936–1960* (New York: Vintage Books, 1966), p. 2.

1. Voices and Echoes in the Public Interest Marketplace: The Development of the Public Interest Sector

1. Ralph Nader's first target of consumer activism was the American automobile industry and, in particular, the manufacturers of the now extinct Corvair. He presented his case in *Unsafe at Any Speed: The Designed-In Dangers of the American Automobile* (New York: Grossman, 1965).

2. For contemporary conceptualizations of political activism, see Sidney Verba, Kay Lehman Schlozman, and Henry E. Brady, *Voice and Equality: Civic Voluntarism in American Politics* (Cambridge, Mass.: Harvard University Press, 1995); Steven J. Rosenstone and John Mark Hansen, *Mobilization, Participation, and Democracy in America* (New York: Macmillan, 1993); and Robert H. Salisbury et al., "Triangles, Networks, and Hollow Cores: The Complex Geometry of Washington Interest Representation," in Mark P. Petracca, ed., *The Politics of Interests: Interest Groups Transformed*, pp. 130–149 (Boulder: Westview Press, 1992).

3. Development directors and direct mail list brokers interviewed by the author present this general profile based on their analysis of membership attributes and opinions.

4. Quoted in Ronald G. Shaiko, "Greenpeace U.S.A.: Something Old, New, Borrowed," *The Annals* 528 (Summer 1993): 93.

5. Ralph Nader, "Foreword,"in Foundation for Public Affairs, *Public Interest Profiles, 1992–1993*, p. xi, opening quotation found on pp. xv–xvi (Washington, D.C.: Congressional Quarterly Press, 1992).

6. David Cohen, "The 1990s and the Public Interest Movement," in Leslie Swift-Rosenzweig, ed., *Public Interest Profiles, 1988–1989*, p. xv (Washington, D.C.:

Congressional Quarterly Press, 1988); see also Seymour Martin Lipset, *American Exceptionalism: A Double-Edged Sword* (New York: Norton, 1996).

7. For subsequent research that builds upon the approach presented in the following chapters, the author has expanded his elite interviewing to include more than 200 lobbyists representing a diverse range of economic, social, and political interests. The conclusions reached in the preceding paragraph are based on these interviews. See also Meredith J. Levin, ed., *Lobbying Techniques for the '90s: Strategies, Coalitions, and Grass-Roots Campaigns* (Washington, D.C.: Congressional Quarterly, 1991): 1–78. Both academics and journalists have emphasized these two methods of influence; for analyses of coalition building see, e.g., Kevin Hula, "Rounding Up the Usual Suspects: Forging Interest Group Coalitions in Washington," in Allan J. Cigler and Burdett A. Loomis, eds., *Interest Group Politics*, 4th ed., pp. 239–258 (Washington, D.C.: Congressional Quarterly Press, 1995); Loree Bykerk and Ardith Maney, "Consumer Groups and Coalition Politics on Capitol Hill," in Cigler and Loomis, eds., *Interest Group Politics*, 4th ed., pp. 259–280; and Colton Campbell and Roger Davidson, "Coalition Building in Congress," in Paul S. Herrnson, Ronald G. Shaiko, and Clyde Wilcox, eds., *The Interest Group Connection: Electioneering, Lobbying, and Policymaking in Washington*, pp. 116–136 (Chatham, N.J.: Chatham House, 1998). For analyses of grassroots lobbying efforts, see, e.g., William P. Browne, "Organized Interests, Grassroots Confidants, and Congress," in Cigler and Loomis, eds., *Interest Group Politics*, 4th ed., pp. 281–298; James G. Gimpel, "Grassroots Organizations and Equilibrium Cycles in Group Mobilization and Access," in Herrnson, Shaiko, and Wilcox, eds., *The Interest Group Connection*, pp. 100–115. See also Peter H. Stone, "Learning from Nader," *National Journal*, June 11, 1994, pp. 1342–1344.

8. See Ronald G. Shaiko, "Lobby Reform: Curing the Mischiefs of Factions?" in James A. Thurber and Roger H. Davidson, eds., *Remaking Congress: Change and Stability in the 1990s*, pp. 156–173 (Washington, D.C.: Congressional Quarterly Press, 1995).

9. Cohen, "The 1990s and the Public Interest Movement," p. xi.

10. Verba, Schlozman, and Brady, *Voice and Equality*, pp. 49–96.

11. The size of public interest sector derived from Ronald J. Hrebenar and Ruth K. Scott, *Interest Group Politics in America* (Englewood Cliffs, N.J.: Prentice-Hall, 1990), p. 11; and Jonathan Rauch, *Demosclerosis: The Silent Killer of American Government* (New York: Times Books, 1994).

12. Philip A. Mundo, *Interest Groups: Cases and Characteristics* (Chicago: Nelson-Hall, 1992), p. 165.

13. See Allan J. Cigler and Anthony J. Nownes, "Public Interest Entrepreneurs and Group Process," in Cigler and Loomis, eds. *Interest Group Politics*, 4th ed., pp. 77–100; Anthony J. Nownes and Grant Neeley, "Public Interest Group Entrepreneurship: Disturbances, Patronage, and Personal Sacrifices," presented at the Annual Meeting of the Midwest Political Science Association, April 6–8, 1995, Chicago; Douglas Imig, "Resource Mobilization and Survival Tactics of Poverty Advo-

cacy Groups," *Western Political Quarterly* 45 (June 1992): 501–520; John Mark Hansen, "The Political Economy of Group Membership," *American Political Science Review* 79 (March 1985): 79–96; Jack L. Walker, "The Origins and Maintenance of Interest Groups in America," *American Political Science Review* 77 (June 1983): 390–406. For a more focused assessment of the environmental sector, see Christopher J. Bosso, "The Color of Money: Environmental Groups and the Pathologies of Fund Raising," in Cigler and Loomis, eds., *Interest Group Politics*, 4th ed., pp. 101–130.

14. Kay Lehman Schlozman and John T. Tierney, *Organized Interests and American Democracy* (New York: Harper and Row, 1986), p. 388.

15. For a complementary assessment of the larger interest group sector, see Allan J. Cigler and Burdett A. Loomis, "Contemporary Interest Group Politics: More than 'More of the Same'," in Cigler and Loomis. eds., *Interest Group Politics*, 4th ed., pp. 393–406.

16. See, e.g., David B. Truman, *The Governmental Process*, 2nd ed., (New York: Knopf, 1971; original edition, 1951).

17. Robert H. Salisbury, "An Exchange Theory of Interest Groups," *Midwest Journal of Political Science* 13 (1969): 1–32. See also Norman Frolich, Joe Oppenheimer, and Oran Young, *Political Leadership and Collective Goods* (Princeton, N.J.: Princeton University Press, 1971).

18. Jeffrey M. Berry, *Lobbying for the People: The Political Behavior of Public Interest Groups* (Princeton, N.J.: Princeton University Press, 1977), p. 24; see also Berry, "On the Origins of Public Interest Groups: A Test of Two Theories," *Polity* 10 (1978): 379–397.

19. Andrew W. McFarland, *Public Interest Lobbies: Decision Making on Energy* (Washington, D.C.: American Enterprise Institute, 1976), chapter 1; see also McFarland, *Common Cause: Lobbying in the Public Interest* (Chatham, N.J.: Chatham House, 1984), chapter 2.

20. McFarland, *Public Interest Lobbies*, pp. 4–5.

21. E. E. Schattschneider, *The Semi-Sovereign People* (New York: Holt, Rinehart, and Winston, 1960); see also McFarland, "Public Interest Lobbies versus Minority Faction," in Allan J. Cigler and Burdett A. Loomis, eds., *Interest Group Politics*, pp. 324–353 (Washington, D.C.: Congressional Quarterly Press, 1983).

22. McFarland, *Common Cause: Lobbying in the Public Interest*, p. 23; for a more recent assessment of Common Cause, see Lawrence S. Rothenberg, *Linking Citizens to Government: Interest Group Politics at Common Cause* (New York: Cambridge University Press, 1992).

23. McFarland, "Interest Groups and Political Time: Cycles in America," *British Journal of Political Science* 21 (July 1991): 257–284. For a synthesis of social movement perspectives, see McFarland, "Social Movements and Theories of American Politics," in Anne N. Costain and Andrew S. McFarland, eds., *Social Movements and American Political Institutions*, pp. 7–19 (Lanham, Md.: Rowman and Littlefield, 1998).

24. See Robert H. Wiebe, *The Search for Order, 1877–1920* (New York: Hill and Wang, 1967); Richard Hofstadter, *The Age of Reform* (New York: Vintage Books, 1955); Alexis de Tocqueville, *Democracy in America* (New York: Mentor Books, 1956). Cf. Mancur Olson, *The Logic of Collective Action* (Cambridge: Harvard University Press, 1965; repr. 1971).

25. The Berry data (N = 83) to be presented in this chapter were drawn from Jeffrey M. Berry, *Lobbying for the People*, chapters 2 and 3. The Foundation for Public Affairs data (N = 250) were drawn from the Foundation's publications; see Douglas J. Bergner, ed., *Public Interest Profiles, 1986–1987*, 5th ed. (Washington, D.C.: Foundation for Public Affairs, 1986); Leslie Swift-Rosenzweig, ed., *Public Interest Profiles, 1992–1993*, 7th ed. (Washington, D.C.: Congressional Quarterly Press, 1992); and Paul McClure, ed., *Public Interest Profiles, 1996–1997*, 8th ed. (Washington, D.C.: Congressional Quarterly Press, 1996). For each group presented, organizational variables were coded by the author from the raw data presented in each profile. The total number of cases presented for each year: 221 (1985 data), 197 (1990 data), and 233 (1995 data) reflect missing data for some organizations as well as the elimination of several "Business/Economic" groups, due to their 501(c)(6) tax status as "Business Organizations." For additional discussion of the Foundation for Public Affairs data sets, see Ronald G. Shaiko, "More Bang for the Buck: The New Era of Full-Service Public Interest Organizations," in Allan J. Cigler and Burdett A. Loomis, eds., *Interest Group Politics*, 3rd ed., pp. 109–130 (Washington, D.C.: Congressional Quarterly Press, 1991).

26. Bergner, ed., *Public Interest Profiles* (1986), p. iii; Swift-Rosenzweig, ed., *Public Interest Profiles* (1992), p. xvii; McClure, ed., *Public Interest Profiles* (1996), p. xi. In the tables that follow, the organizational data are reported for the year in which they were collected rather than for the date of publication of the volume from which they were taken.

27. Interestingly, the Vietnam issue, recast by Vietnam veterans and their supporters in the late 1970s and early 1980s, has managed to reclaim a position on the public agenda and has resulted in the remobilization and organization of citizens concerned with the plight of those who survived the war.

28. The budget figures presented by Berry were converted to 1990 dollars by multiplying by a factor equal to the 1990 Consumer Price Index (2.391) divided by the 1972 Consumer Price Index (0.766), or roughly multiplying by 3.0 (actual conversion factor is 3.12). Therefore, a budget of $50,000 in 1972 is roughly equal to $150,000 in 1990. The year 1990 was chosen because it represents the midpoint between the three FPA data sets and the conversion factor for that year allows the Berry data to be clearly translated.

29. Reporting of organizational support in the 1995 FPA data was modified from earlier volumes. In many cases, levels of foundation and corporate support were combined. As a result only the findings from the 1985 and 1990 FPA data sets are reported.

30. McFarland, *Public Interest Lobbies*, p. 37.

31. Controlling for organizations with no members or organizational memberships, the percentages of organizations with memberships greater than 200,000 in each sample are: Berry 1972, 12.8 percent; FPA 1985, 34.0 percent; FPA 1990, 35.8 percent; and FPA 1995, 36.3 percent.

32. Most public interest organizations are classified as either 501(c)(3) tax-exempt, tax-deductible "charitable" organizations or 501(c)(4) tax-exempt "social welfare" organizations. However, other organizations included in the interest group community fall into several other IRS categories: Veteran's Organizations, 501(c)(19); Labor and Agricultural Organizations, 501(c)(5); Business Organizations, 501(c)(6); and Employees' Associations, 501(c)(4), (9), and (17). See United States General Accounting Office, "Tax-Exempt Organizations: Information on Selected Types of Organizations," (February 1995), GAO/GGD-95-84BR; Internal Revenue Service, "Tax-Exempt Status For Your Organization," Publication 557 (Washington, D.C.: Department of the Treasury, rev. July 1985). For an extensive analysis of the effects of tax status on the political activities of 501(c)(3) and 501(c)(4) organizations, see Shaiko, "More Bang for the Buck."

33. Paul E. Treusch and Norman A. Sugarman, *Tax-Exempt Charitable Organizations,* 2nd ed. (Philadelphia: American Law Institute–American Bar Association, 1983). The alternative test provided in the 1976 Tax Reform Act allows for a sliding scale that permits an organization to spend 20 percent of its first $500,000 on direct lobbying efforts, 15 percent of the next $500,000, and 5 percent of its remaining budget. A total of 25 percent of these expenditures may be used on indirect lobbying efforts (citizen mobilizations, grass roots lobbying). See Bruce R. Hopkins, *The Law of Tax Exempt Organizations,* 2nd ed. (Washington, D.C.: Lerner, 1977), p. 149; and Burton A. Weisbrod, *The Nonprofit Economy* (Cambridge, Mass.: Harvard University Press, 1988), p. 120. For an applied analysis of nonprofit sector lobbying, see Bob Smucker, *The Nonprofit Lobbying Guide: Advocating Your Cause—and Getting Results* (San Francisco: Jossey-Bass, 1991).

34. William J. Lehrfeld, quoted in John Riley, "Tax-Exempt Foundations: What is Legal?" *The National Law Journal* 9 (24) (February 23, 1987): 8.

35. Environmental Action, "1985 Annual Report: Environmental Action/Environmental Action Foundation," (Washington, D.C.: Environmental Action, 1985).

36. John Wark and Gary Marx, "Faith, Hope, and Chicanery," *Washington Monthly* 19 (January 1987): 29.

37. Edward T. Pound, Gary Cohen, and Penny Loeb, "Tax Exempt!" *U.S. News and World Report,* October 2, 1995, 36–39, 42–44, 46, 51. For a more focused analysis of religious nonprofit organizations, see Stephen V. Monsma, *When the Sacred and the Secular Mix: Religious Nonprofit Organizations and Public Money* (Lanham, Md.: Rowman and Littlefield, 1996).

38. Richard C. Sachs, "The Lobbying Disclosure Act of 1995: A Brief Description," *CRS Report to Congress 96–29GOV,* (January 4, 1996), pp. 1–6.

39. Shaiko, "Lobby Reform: Curing the Mischiefs of Factions?"

40. The AARP, a 501(c)(4) organization, has responded to the new law by re-

ceiving all government grants and contracts through its 501(c)(3) foundation, thereby allowing the parent organization to continue lobbying.

41. See "The Istook-McIntosh-Ehrlich Proposal: Hearing before the Subcommittee on National Economic Growth, Natural Resources, and Regulatory Affairs of the Committee on Government Reform and Oversight," U.S. House of Representatives, 104th Cong., 1st sess., September 28, 1995.

42. Both Riley and Browner, however, have former ties to institutional Washington. Riley served as legal counsel to former South Carolina Senator Olin Johnston while Browner was legislative director for Senator Al Gore (D-TN) as well as chief environmental aide to Senator Lawton Chiles (D-FL). In addition to Browner, there are several Clinton subcabinet appointees with ties to national environmental organizations, including EPA Assistant Administrator for Policy and Planning David Gardiner, formerly the director of the Washington office of the Sierra Club; Assistant Secretary of Interior George Frampton, formerly president of The Wilderness Society; and Jim Baca, director of the Bureau of Land Management, formerly a Wilderness Society board member.

43. Deputy Secretary of State Clifton Wharton Jr. served with Christopher and Rivlin on the board of the Council on Foreign Relations. He also served with Shalala on the board of the Committee for Economic Development. In 1996 Rivlin left OMB to become the vice-chairperson of the Federal Reserve Board.

44. For additional analysis of the representation of women in public interest leadership roles, see Ronald G. Shaiko, "Female Participation in Public Interest Nonprofit Governance: Yet Another Glass Ceiling?" *Nonprofit and Voluntary Sector Quarterly* 25 (3) (September 1996): 302–320.

45. Mancur Olson, Jr., *Logic of Collective Action*.

46. See, e.g., Terry M. Moe, *The Organization of Interests*: Incentives and the Internal Dynamics of Political Interest Groups (Chicago: University of Chicago Press, 1980); Moe, "A Calculus of Group Membership," *American Journal of Political Science* 24 (1980): 593–632; David Knoke, *Organizing for Collective Action* (New York: Aldine de Gruyter, 1990); Knoke and James R. Wood, *Organized for Action* (New Brunswick, N.J.: Rutgers University Press, 1981); Knoke, "Commitment and Detachment in Voluntary Associations," *American Sociological Review* 46 (1981): 141–158. See also James Q. Wilson, *Political Organizations* (New York: Basic Books, 1973); and William A. Gamson, *The Strategy of Social Protest* (Homewood, Ill.: Dorsey Press, 1975). Besides Moe and Knoke, numerous scholars have shed important light on the issue of organizational maintenance and political representation in the public interest sector. The bibliography presented at the end of this volume includes a wide variety of books and articles that contributed to the theoretical and analytical framework of my research.

47. See John D. McCarthy and Mayer N. Zald, "Resource Mobilization and Social Movements: A Partial Theory," *American Journal of Sociology* 82 (1977): 1212–1241; Theodore J. Lowi, *The Politics of Disorder* (New York: Basic Books, 1971); McFarland, *Public Interest Lobbies*.

48. McCarthy and Zald, "Resource Mobilization and Social Movements," p. 1213. The most comprehensive statements of this perspective are offered by Anthony Oberschall, *Social Conflicts and Social Movements* (Englewood Cliffs, N.J.: Prentice-Hall, 1973); McCarthy and Zald, *The Trend in Social Movements in America: Professionalization and Resource Mobilization* (Morristown, N.J.: General Learning Press, 1973); and Charles Tilly, *From Mobilization to Revolution* (Reading, Mass.: Addison-Wesley, 1978). Two more recent volumes, edited by McCarthy and Zald, provide a variety of applications of the resource mobilization approach: *The Dynamics of Social Movements: Resource Mobilization, Social Control, and Tactics* (Lanham, Md.: University Press of America, 1988); and *Social Movements in an Organizational Society* (New Brunswick, N.J.: Transaction Books, 1987). See also, Sidney Tarrow, *Power in Movement: Social Movement, Collective Action, and Politics* (New York: Cambridge University Press, 1994).

49. Laura R. Woliver, *From Outrage to Action: The Politics of Grassroots Dissent* (Urbana: University of Illinois Press, 1993); Jeffrey M. Berry, Kent E. Portney, and Ken Thomson, *The Rebirth of Urban Democracy* (Washington, D.C.: Brookings Institution, 1993).

50. Keith Schneider, "Pushed and Pulled, Environment Inc. Is on the Defensive," *New York Times*, March 29, 1992, pp. 1, 4. For a more detailed assessment, see Mark Dowie, *Losing Ground: American Environmentalism at the Close of the Twentieth Century* (Cambridge, Mass.: MIT Press, 1995); see also Barry G. Rabe, *Beyond Nimby: Hazardous Waste Siting in Canada and the United States* (Washington, D.C.: Brookings Institution, 1994).

51. See, e.g., Paul Wapner, *Environmental Activism and World Civic Politics,* (Albany: State University of New York Press, 1996); Gareth Porter and Janet Welsh Brown, *Global Environmental Politics*, 2nd ed. (Boulder: Westview Press, 1995); and Ken Conca, Michael Alberty, and Geoffrey D. Dabelko, eds., *Green Planet Blues: Environmental Politics from Stockholm to Rio* (Boulder: Westview Press, 1995).

52. Jo Freeman, *The Politics of Women's Liberation* (New York: David McKay, 1975), p. 100.

53. Moe, *The Organization of Interests*, pp. 259–260.

54. Cohen, "The 1990s and the Public Interest Movement," p. xv.

55. Hannah F. Pitkin, *The Concept of Representation* (Berkeley: University of California Press, 1967), pp. 221–222.

56. Norman R. Luttberg makes a distinction between coercive and noncoercive models of representation, see Luttberg, *Public Opinion and Public Policy* (Homewood, Ill.: Dorsey Press, 1974); see also Albert O. Hirschman, *Exit, Voice, and Loyalty* (Cambridge: Harvard University Press, 1970).

57. Pitkin, *The Concept of Representation*, pp. 112–143. The concept of virtual representation is derived from the writings of Edmund Burke. Burke believed that "the kings, the lords, and the judges" were better equipped to decide for the masses what was best for them. As a member of the British parliament in the late 1700s, Burke argued that an elected representative "owes you (the constituency), not his

industry only, but his judgment; and he betrays, instead of serving you, if he sacrifices it to your opinion." Edmund Burke, "On Election to Parliament," reprinted in Diane Ravitch and Abigail Thernstrom, eds., *The Democracy Reader* (New York: HarperCollins, 1992), p. 50. Virtual representation, acting for the interests of others, is implicitly elitist in its application.

2. From Social Movement to Public Interest Organizations: The Organizational Transformation of Environmentalism in the United States

1. In comparison with the civil rights movement, the consumer movement, the women's movement or, more recently, the gay and lesbian rights movement, the animal rights movement, or even the property rights movement, the organizational transformation that has taken place within the environmental movement is unparalleled in American social movement mobilization and activism.

2. Excellent historical analyses include: Philip Shabecoff, *A Fierce Green Fire: The American Environmental Movement* (New York: Hill and Wang, 1993); Stephen Fox, *The American Conservation Movement* (Madison: University of Wisconsin Press, 1985); Fox, *John Muir and His Legacy* (Boston: Little, Brown, 1981); and Samuel P. Hays, *Beauty, Health, and Permanence: Environmental Politics in the United States, 1955–1985* (New York: Cambridge University Press, 1987). Earlier writings on the history of the movement include: Grant McConnell, "The Conservation Movement: Past and Present," *Western Political Quarterly* 7 (1954): 463–578; Michael McCloskey, "Wilderness at the Crossroads, 1945–1970," *Pacific Historical Review* 41 (1972): 346–361; see also Allan Schnaiberg, *The Environment: From Surplus to Scarcity* (New York: Oxford University Press, 1980). Additional analyses of the contemporary environmental movement include: Christopher J. Bosso, *Pesticides and Policies: The Life Cycle of a Public Issue* (Pittsburgh: University of Pittsburgh Press, 1987); Bosso, "Adaptation and Change in the Environmental Movement," in Allan J. Cigler and Burdett A. Loomis, eds., *Interest Group Politics*, 3rd ed. (Washington, D.C.: Congressional Quarterly Press, 1991), 151–176; Robert C. Paehlke, *Environmentalism and the Future of Progressive Policies* (New Haven: Yale University Press, 1989); Victor B. Scheffer, *The Shaping of Environmentalism in America* (Seattle: University of Washington Press, 1991); Donald Snow, *Inside the Environmental Movement: Meeting the Leadership Challenge* (Washington, D.C.: Island Press, 1992); Bosso, "After the Movement: Environmental Activism in the 1990s," in Norman J. Vig and Michael E. Kraft, eds., *Environmental Policy in the 1990s* (Washington, D.C.: Congressional Quarterly Press, 1994); Kirkpatrick Sale, *The Green Revolution: The American Environmental Movement, 1962–1992* (New York: Hill and Wang, 1993); Robert Gottlieb, *Forcing the Spring: The Transformation of the American Environmental Movement* (Washington, D.C.: Island Press, 1993); and most recently, Mark Dowie, *Losing Ground: American Environmentalism at the Close of the Twentieth Century* (Cambridge, Mass.: MIT Press, 1995).

3. Grant McConnell, "The Conservation Movement," p. 463.

4. Michael P. Cohen, "Origins and Early Outings," and Douglas Scott, "Con-

servation and the Sierra Club," in Patrick Carr, ed., *The Sierra Club: A Guide* (San Francisco: Sierra Club Books, 1989), pp. 9–17, 18–37.

5. See, e.g., Robert C. Mitchell, "From Conservation to Environmental Movement: The Development of Modern Environmental Lobbies," in Michael J. Lacey, ed., *Government and Environmental Politics: Essays on Historical Developments Since World War II* (Baltimore: Johns Hopkins University Press, 1991), pp. 81–114; Samuel P. Hays, *Beauty, Health, and Permanence,* pp. 1–12; and Hays, "Three Decades of Environmental Politics: The Historical Context," in Lacey, ed., *Government and Environmental Politics,* pp. 19–80. For a fascinating view of the development of the American environmental movement from 1866 to 1992 in time line form, see Kirkpatrick Sale, *The Green Revolution,* pp. xi–xx.

6. Robert Mitchell, "From Conservation to Environmental Movement," p. 82.

7. See e.g., Rachel Carson, *Silent Spring* (Boston: Houghton Mifflin, 1962); Barry Commoner, *Science and Survival* (New York: Ballantine Books, 1963); Paul R. Ehrlich, *The Population Bomb* (New York: Ballantine Books, 1968); Garrett Hardin, "The Tragedy of the Commons," *Science* 162 (1968): 1243–1255; Dirck Van Sickle, *The Ecological Citizen* (New York: Harper and Row, 1971). In addition to these new commentaries, the works of earlier conservationists such as Aldo Leopold, John Muir, and Roderick Nash were reissued during this period.

8. See e.g., Riley E. Dunlap and Angela G. Mertig, "The Evolution of the U.S. Environmental Movement from 1970 to 1990: An Overview," and Robert C. Mitchell, Angela G. Mertig, and Riley E. Dunlap, "Twenty Years of Environmental Mobilization: Trends Among National Environmental Organizations," in Riley E. Dunlap and Angela G. Mertig, eds., *American Environmentalism: The U.S. Environmental Movement, 1970–1990* (Philadelphia: Taylor and Francis, 1992), pp. 1–10, 11–26.

9. See Jack L. Walker Jr., "The Origins and Maintenance of Interest Groups in America," *American Political Science Review* 77 (1983): 390–406. In addition to the organizations mentioned above, the Union of Concerned Scientists (1969), the League of Conservation Voters (1970) and Greenpeace (1971) were formed during this period, as was the Ralph Nader consumer organization Public Citizen (1971). Direct action groups began appearing in the late 1970s, with the Sea Shepherd Conservation Society forming in 1977 and Earth First! in 1980.

10. Grant P. Thompson, "The Environmental Movement Goes to Business School," *Environment* 27 (May 1985): 9–10. For additional commentaries on the professionalization of the environmental movement, see Bill Keller, "Environmental Movement Checks Its Pulse and Finds Obituaries are Premature," *Congressional Quarterly Weekly Report,* January 31, 1981, pp. 211–216; Rochelle Stanfield, "Environmental Lobby's Changing of the Guard Is Part of Movement's Evolution," *National Journal,* June 5, 1985, pp. 1350–1353; and Snow, *Inside the Environmental Movement.*

11. David Roe, *Dynamos and Virgins* (New York: Random House, 1984), p. 186.

12. Peter Borrelli, "Environmentalism at a Crossroads," in Peter Borrelli, ed.,

Crossroads: Environmental Priorities for the Future (Washington, D.C.: Island Press, 1988), p. 10. An analysis of the ten largest environmental organizations conducted by Peter Overberg and Linda Kanamine for *USA Today* found that by 1990 the combined memberships of the top ten groups alone amounted to 8.2 million; their combined budgets in 1992 totaled more than $300 million. See Overberg and Kanamine, "Green But Not Growing," *USA Today*, October 19, 1994, p. 8A. See also Kirkpatrick Sale, *The Green Revolution*, pp. 77–81; and Sale, "The U.S. Green Movement Today," *The Nation*, July 19, 1993, pp. 92–96. In broadest terms, according to the American Association of Fund-Raising Counsel, public and private, corporate, and foundation donations to environmental and wildlife causes totaled $3.2 billion in 1993; see Daniel Borochoff, "Overview: How Americans Give," *AIP Charity Rating Guide and Watchdog Report* (Summer/Fall 1994): 15.

13. Organizations such as the American Civil Liberties Union and People for the American Way used Edwin Meese, attorney general in the Reagan administration, in a similar fashion with great success.

14. Howard Youth, "Boom Time for Environmental Groups," *WorldWatch* (November/December 1989): 33.

15. Between 1992 and 1993, memberships of these three organizations grew by 25 percent, 13 percent, and 11 percent, respectively; see Wendy Taylor, "Enviro Groups: Success at EDF, TNC, and NPCA No Accident," *Greenwire*, December 16, 1994, pp. 1–5.

16. Philanthropic Research, Inc., *GuideStar Directory of American Charities: 1996 Index* (Williamsburg: Philanthropic Research, 1996). From information provided by 501(c)(3) organizations on their IRS 990 forms, Philanthropic Research, Inc., has created a directory of 35,000 organizations. Their 1995 IRS data on national environmental organizations show combined budgets for 501(c)(3) organizations to be more than $650 million. With the budgets of 501(c)(4) organizations, such as the Sierra Club and Clean Water Action, included (see table 2.5), the total budget figure surpasses $700 million.

17. Brad Knickerbocker, "Nation's Green Advocates See Their Groups on Critical List," *Christian Science Monitor*, October 17, 1994, p. 1.

18. Everett Carll Ladd and Karlyn H. Bowman, *Attitudes Toward the Environment: Twenty-Five Years After Earth Day*, (Washington, D.C.: AEI Press, 1995), p. 7.

19. Ladd and Bowman, *Attitudes Toward the Environment*, pp. 45–46. VNS pollsters did include questions regarding the environment on nine statewide polls in the western part of the country in 1994. Eight of the nine states elected Republican governors. In eight of the nine states a majority of voters felt that President Clinton's environmental and land use policies had hurt their states; see Ladd and Bowman, p 45.

20. The Clinton/Gore campaign was successful in moving Medicare and education into the top five issues identified as most important in the 1996 VNS exit polls as Medicare/social security followed economy/jobs as the second most important

issue with 15 percent and education tied for third with the federal budget at 12 percent.

21. For an analysis of the relationships between news media coverage and public opinion formation, see Benjamin I. Page and Robert Y. Shapiro, *The Rational Public: Fifty Years of Trends in Americans' Policy Preferences* (Chicago: University of Chicago Press, 1992), pp. 339–354.

22. Anthony Downs, "Up and Down with Ecology: The Issue Attention Cycle," *Public Interest* 94 (Summer 1972): 28–38.

23. The data presented in figure 2.1 are drawn from the *New York Times* index of news stories from 1968 through 1996 under the subject heading of "environment" and include both front-page and nonfront-page stories. Environmental news coverage by the *New York Times* staff provides a representative measure of national media coverage of the environment, with several journalists serving as environmental reporters throughout the time period analyzed. Philip Shabecoff, author of *A Fierce Green Fire: The American Environmental Movement,* served as chief environmental reporter for the *New York Times* for almost fifteen years; he is currently executive publisher of *Greenwire,* a daily environmental news service. For a slightly different view of media salience of environmental issues, see Timothy A. Huelskamp, "Congressional Change: Committees on Agriculture in the U.S. Congress" (Ph.D. diss., The American University, 1995), p. 102.

24. Bernard Cohen, *The Press and Foreign Policy* (Princeton, N.J.: Princeton University Press, 1963); Donald L. Shaw and Maxwell E. McCombs, *The Emergence of American Political Issues: The Agenda Setting Function of the Press* (St. Paul: West Publishing, 1977); Martin Linsky, *Impact: How the Press Affects Federal Policymaking* (New York: Norton, 1986); Shanto Iyengar and Donald Kinder, *News that Matters: Television and American Opinion* (Chicago: University of Chicago Press, 1987). For an excellent assessment of the role of the media in the electoral context, see Thomas E. Patterson, *The Mass Media Election: How Americans Choose Their Presidents* (New York: Praeger, 1980); see also Thomas E. Patterson, *Out of Order* (New York: Knopf, 1993) for a critique of the media's capacity to serve as an agenda setter in American electoral politics.

25. See Richard E. Cohen, *Washington at Work: Back Rooms and Clean Air* (New York: MacMillan, 1992), for an assessment of the political process that produced the Clean Air Act Amendments in 1990.

26. Members of the Republican leadership in the House of Representatives have demonstrated a lack of support for environmental issues through their voting records. The League of Conservation Voters (LCV) rates members of Congress annually on their environmental voting records by scoring a series of key environmental votes cast in each session of Congress. Prior to taking control of the House and Senate in the 104th Congress, the top nine members of the Republican leadership team received an average rating score of 9 from the LCV in 1994 (on a scale of 0–100, with 100 being the strongest environmental support). House Speaker Newt Gingrich received a rating of 0; Majority Leader Dick Armey received a 4, as did

Majority Whip Tom DeLay, chief proponent of the Republican regulatory reform agenda. See Sarah Anderson, Paul Brotherton, and Peter L. Kelley, eds., *The Score-card* (Washington, D.C.: League of Conservation Voters, 1994).

27. Ladd and Bowman, *Attitudes Toward the Environment*, p. 51; see also George Pettinico, "The Public Interest Paradox," *Sierra* 80 (November/December 1995): 28, 30–31.

28. For analyses of direct action organizations in the environmental movement, see Ronald G. Shaiko, "Greenpeace U.S.A.: Something Old, New, Borrowed," *The Annals* 528 (July 1993): 88–100; Martha F. Lee, *Earth First! Environmental Apocalypse* (Syracuse, N.Y.: Syracuse University Press, 1995); Dick Russell, "The Monkey Wrenchers," in Peter Borrelli, ed., *Crossroads: Environmental Priorities for the Future*, 27–48; and Christopher Manes, *Green Rage: Radical Environmentalism and the Unmaking of Civilization* (Boston: Little, Brown, 1990).

29. Shaiko, "Greenpeace, U.S.A.: Something Old, New, Borrowed," pp. 88–100.

30. Compare the findings in tables 1.1–1.4 with the organizational attributes of environmental organizations presented in table 2.5.

31. Each of the case studies is based on materials made available by the organizations.

32. The name, Sierra Club and the organization's slogan, "One Earth, One Chance," are registered trademarks of the Sierra Club.

33. Ronald G. Shaiko, "More Bang for the Buck: The New Era of Full-Service Public Interest Organizations," in Allan J. Cigler and Burdett A. Loomis, eds., *Interest Group Politics*, 3d ed., pp. 109–130.

34. Recall from chapter 1 that the Sierra Club originally had 501(c)(3) tax status but relinquished it in 1966 due to excessive grassroots lobbying efforts.

35. Sierra Club PAC, the second largest environmental PAC, contributed $406,631 to congressional candidates; 98 percent of their contributions went to Democrats. The largest environmental PAC in the 1994 cycle was the League of Conservation Voters PAC, with contributions of $800,207, 96 percent of which went to Democrats. LCV PAC ranked second in contributions among all ideological, single-issue PACs; the National Rifle Association PAC was the largest spender, with $1,853,038 in direct contributions to candidates. See Larry Makinson, *The Price of Admission: Campaign Spending in the 1994 Elections* (Washington, D.C.: Center for Responsive Politics, 1995), p. 235.

36. B. J. Bergmann, "Majority Rules, And It's Green," *Sierra*, January/February 1997, pp. 50, 52–53.

37. This figure includes the sales of more than 27,000 copies of *Sierra* on average per issue by newsstands and bookstores across the country.

38. National Wildlife Productions, Inc., and International Wildlife Incorporated are also housed in the national headquarters.

39. Like the Sierra Club and The Wilderness Society, the National Wildlife Federation also cut its staff significantly as a result of budget shortfalls and declining memberships in the early 1990s. At one point in the 1990s, the Federation had more than 800 employees.

40. In 1994 the National Wildlife Federation reported total revenues of $101.4 million, with total budget expenditures of $96.3 million, resulting in a budget surplus of more than $4 million, after covering a Real Estate Limited Partnership loss of $900,000. Associate members of the National Wildlife Federation receive *National Wildlife* magazine and, for an additional fee, may also get *International Wildlife* magazine. The Federation's four magazines have a combined circulation of 1.7 million copies.

41. Not surprisingly, the bulk of the organizational expenses were generated through the production and promotion of these goods and services. Approximately $55 million was spent in 1994 on the production and promotion of *Ranger Rick, Your Big Backyard,* and on nature education materials.

3. Growing Pains: Leadership Challenges in Contemporary Environmental Organizations

1. See Stephen Fox, *John Muir and His Legacy* (Boston: Little, Brown, 1981); and Fox, *The American Conservation Movement* (Madison: University of Wisconsin Press, 1985).

2. Quoted in Donald Snow, *Inside the Environmental Movement: Meeting the Leadership Challenge* (Washington, D.C.: Island Press, 1992), pp. 150–151.

3. Daniel Borochoff, "Overview: How Americans Give," *AIP Charity Rating Guide and Watchdog Report* (Summer/Fall 1994): 15.

4. Ronald G. Shaiko, "Female Participation in Public Interest Nonprofit Governance: Yet Another Glass Ceiling?" *Nonprofit and Voluntary Sector Quarterly* 25 (3) (September 1996): 302–320.

5. See Robert H. Salisbury, "An Exchange Theory of Interest Groups," *Midwest Journal of Political Science* 13 (July 1969): 1–32; Jeffrey M. Berry, "On the Origins of Public Interest Groups: A Test of Two Theories," *Polity* 10 (Spring 1978): 379–397.

6. Quoted in Borrelli, "Environmentalism at a Crossroads," p. 17.

7. The Group of Ten included the Environmental Defense Fund, the Izaak Walton League, the National Audubon Society, the National Wildlife Federation, the Natural Resources Defense Council, the Sierra Club, The Wilderness Society, the Environmental Policy Institute, the Friends of the Earth, and the National Parks and Conservation Association.

8. The Green Group includes the original members of the Group of Ten as well as the Sierra Club Legal Defense Fund, the World Wildlife Fund, the Children's Defense Fund, the National Toxics Campaign, the Union of Concerned Scientists, the Native American Rights Fund, Planned Parenthood, the Population Crisis Committee, and Zero Population Growth, and the Defenders of Wildlife.

9. Matthew Reed Baker, "What Interest Groups Pay Their Leaders," *National Journal,* April 26, 1997, pp. 807–817. The sample of "public interest group" leaders includes seventy-two executive directors or presidents.

10. In comparison with all other interest group sectors, public interest executives receive the lowest salaries. Executive directors of associations in the finance, insurance, and real estate industries, for example, earn more than $350,000 in annual

salaries on average. The highest paid executive director in the Washington interest group community is Jason Berman, chairman of the Recording Industry Association of America, with an annual compensation package of $1,247,864; see Baker, "What Interest Groups Pay Their Leaders," p. 808.

11. Of the 50 highest compensated executive directors in the sample of 413 interest group leaders, each earning between $400,000 and $1.2 million, there is only one women on the list (#18), Nancy Fletcher, president of the Outdoor Advertising Association of America, earning an annual salary of $548,779.

12. In a similar analysis conducted by *National Journal* in 1994, 26 percent of the public interest leaders were women. The wage gap between men and women has closed in the three years between the studies as women in the 1994 study earned an average salary of $80,945 while men earned $126,310. See Peter H. Stone, "Payday!" *National Journal,* December 14, 1994, pp. 2948–2960.

13. John W. Gardner, *On Leadership* (New York: Free Press, 1990), pp. 121–137.

14. James MacGregor Burns, *Leadership* (New York: Harper and Row, 1978), pp. 19–20, 141–254, 257–397.

15. Quoted in Carl Pope, "Earth's Future and Us," *San Francisco Chronicle,* April 22, 1993, p. A21.

16. Prior to the election of Charles McGrady as president of the Sierra Club, the organization was presided over by Adam Werbach. Werbach was first elected president of the Sierra Club by the national board of directors in May 1996 at the age of twenty-three, thereby becoming the youngest president in the history of the organization; see Alex Barnum, "A Fresh Look for Sierra Club," *San Francisco Chronicle,* May 25, 1996, p. A1.

17. Carl Nolte, "Sierra Club Settles on 10% Staff Cut," *San Francisco Chronicle,* November 19, 1994, p. B10.

18. Interview, Michael McCloskey, chairman, board of directors, Sierra Club, December 19, 1985, San Francisco, Sierra Club National Headquarters.

19. Marsha Ginsburg, "Sierra Club to Slash Its Staff," *San Francisco Examiner,* November 18, 1994, p. B4.

20. The Sierra Club experienced budgetary and staff cutbacks in 1973 and 1984, but neither downsizing was as significant as the experience in the early 1990s.

21. Quoted in Borrelli, "Environmentalism at a Crossroads," p. 13.

22. Quoted in Ken Miller, "Founder of Many Green Groups Scorns Their Complacency," *Gannett News Service,* April 20, 1995, p. 3.

23. Frank Clifford, "Environmental Movement Struggles as Clout Fades," *Los Angeles Times,* September 21, 1994, p. A1.

24. Jack Thorndike, "Trouble at the Sierra Club," *The Progressive* 56 (July 1992): 13.

25. Thorndike, "Trouble at the Sierra Club," p. 13.

26. Clifford, "Environmental Movement Struggles as Clout Fades," p. A1.

27. Interview, Michael McCloskey, December 19, 1985.

28. Mark Mardon, "Where Are We, Anyway?" *Sierra,* May/June 1991, pp. 18–19.

29. Borochoff, "Overview: How Americans Give," p. 15; see also Burton A. Weisbrod, *The Nonprofit Economy* (Cambridge, Mass.: Harvard University Press, 1988), pp. 93–102; and Andrew W. Osterland, "War Among the Nonprofits," *Financial World*, September 1, 1994, p. 52.

30. Quoted in Carl Nolte, "Sierra Club Lays Off 20, Puts a Freeze on Travel," *San Francisco Chronicle*, July 13, 1991, p. A15.

31. Borochoff, "Overview: How Americans Give," p. 15.

32. See Albert O. Hirschman, *Shifting Involvements: Private Interests and Public Action* (Princeton, N.J.: Princeton University Press, 1982); and R. Kenneth Godwin and Robert C. Mitchell, "The Impact of Direct Mail on Political Organizations," *Social Science Quarterly* 65 (Fall 1984): 829–839.

33. Carl Pope, "Letter to the Editor: Environmental Slippage Blamed on Media, Congressional Foes," *San Francisco Examiner*, October 14, 1994, p. A22.

34. Margaret Kriz, "The Conquered Coalition," *National Journal*, December 14, 1994, p. 2826.

35. Quoted in Kriz, "The Conquered Coalition, " p. 2826; and in "Sierra Club Puts New Legislative Leaders on Notice," *PR Newswire*, November 19, 1994, p. 3.

36. Quoted in "Sierra Club Puts New Legislative Leaders on Notice, " p. 4.

37. "The ABCs of Activism," *Sierra*, January/February 1996, p. 23.

38. See "Sierra Silliness-Editorial," *Providence-Journal Bulletin*, April 24, 1997, p. 7B; "Sierra Club Youthful Prez's Style Draws Praise, Dissent," *Greenwire*, March 25, 1997, p. 1; and Josh Chetwynd, "Splatter-Casting the Sierra Club," *U.S. News and World Report*, March 31, 1997, p. 37.

39. See Stephen Fox, "We Want No Straddlers," *Wilderness* 48 (1985): 5–19; and Robert C. Mitchell, "From Conservation to Environmental Movement," pp. 98–99.

40. Interview, William A. Turnage, executive director (president), The Wilderness Society, October 25, 1985, Washington, D.C., The Wilderness Society National Headquarters.

41. Interview, William Turnage.

42. The Turnage interview took place less than two months before he stepped down as executive director.

43. Interview, William Turnage.

44. Interview, Robert O. Blake, secretary of the board of directors, The Wilderness Society, October 21, 1985, Washington, D.C., International Institute for Environment and Development.

45. James MacGregor Burns, *Leadership* (New York: Harper and Row, 1978), p. 20.

46. Interview, William Turnage.

47. Interview, William Turnage.

48. Interview, William Turnage.

49. H. Watkins and William A. Turnage, "We Still Want No Straddlers," *Wilderness* 48 (1985): 36, 46.

50. "Frampton New Society President," *Wilderness* (Winter 1985): 2.

51. Interview, Sharon Dreyfuss, membership associate, The Wilderness Society, October 17, 1985, Washington, D.C., The Wilderness Society National Headquarters.

52. G. Jon Roush, "Conservation's Hour—Is Leadership Ready?" in Donald Snow, ed., *Voices from the Environmental Movement: Perspectives for a New Era*, pp. 21–40 (Washington, D.C.: Island Press, 1992).

53. Roush, "Conservation's Hour—Is Leadership Ready?" p. 37.

54. Roush, "Conservation's Hour—Is Leadership Ready?" p. 37–39; emphasis in original text.

55. Roush, "Conservation's Hour—Is Leadership Ready?" p. 40.

56. For the distinction between individual and institutional support for national environmental organizations, see Christopher J. Bosso, "The Color of Money: Environmental Groups and the Pathologies of Fund Raising," in Allan J. Cigler and Burdett A. Loomis, eds., *Interest Group Politics*, 4th ed., pp. 106–110 (Washington, D.C.: Congressional Quarterly Press, 1995).

57. Tom Kenworthy, "Wilderness Society President Sold Timber Cut on His Montana Ranch," *Washington Post*, April 7, 1995, p. A28.

58. Baca served less than two years as head of the Bureau of Land Management in the Clinton administration.

59. See Alexander Cockburn, "The Green Betrayers: Major Environmental Groups and the Clinton Administration," *The Nation*, February 6, 1995, p. 157; Cockburn, "Wilderness Society: New Shame," *The Nation*, February 20, 1995, p. 228; Cockburn, "Wilderness Society: The Saga of Shame Continues," *The Nation*, March 6, 1995, p. 300; Cockburn, "Wilderness Society: Saga of Shame (Part MMDVI)," *The Nation*, April 3, 1995, p. 444; Cockburn and Jeffrey St. Clair, "Wilderness Chief in Tree Massacre," *The Nation*, April 24, 1995, pp. 556–559; and Cockburn, "Roush: The Sequel," *The Nation*, May 1, 1995, pp. 588–589.

60. Tom Kenworthy, "Conservationist's Logging Deal Draws Fire," *Los Angeles Times*, April 7, 1995, p. A41.

61. Mary Hanley left The Wilderness Society in 1997 to direct the public affairs office at the Commerce Department.

62. The Wilderness Society, "Meadows Named New President of Wilderness Society," press release, Washington, D.C., September 24, 1996, pp. 1–2.

63. A Lexis-Nexis search for Meadows citations during his first six months in office revealed only three citations—one in a Reuters wire service dispatch and two in an energy trade publication. There were no citations from major press outlets.

64. Interview, Joel Thomas, general counsel, The National Wildlife Federation, March 6, 1986, Vienna, Va., National Wildlife Federation Administrative Headquarters.

65. Interview, John Gottschalk, board of directors, National Wildlife Federation, October 23, 1985, International Association of Fish and Wildlife Agencies Headquarters, Washington, D.C.

66. Interview, Joel Thomas.

67. Interview, John Gottschalk.

68. Paul Overberg and Linda Kanamine, "Green But Not Growing," *USA Today*, October 19, 1994, p. 8A.

69. Overberg and Kanamine, "Green But Not Growing," p. 8A.

70. See, e.g., Michael Kinsley, "The Moral Myopia of Magazines," *Washington Monthly* 7 (September 1975): 7–18; and Lilu Locksley, "The National Geographic: How to Be Non-Profit and Get Rich," *Washington Monthly* 9 (September 1977): 46–48. In the Senate hearings concerning the American Association of Retired Persons conducted in 1995 by Senator Alan Simpson, similar issues were raised in relation to AARP sales of drugs, insurance, and other merchandise and services.

71. Jay D. Hair, "The Earth's Environment: A Legacy in Jeopardy," in Peter Borrelli, ed., *Crossroads: Environmental Priorities for the Future*, p. 201.

72. Eve Pell, "Movements: Buying In," *Mother Jones*, April/May 1990, p. 23.

73. Pell, "Movements: Buying In," p. 23. For assessments of what critics of these linkages between businesses and environmental organizations have labeled "greenwashing," see e.g., Paul Rauber, "Beyond Greenwash," *Sierra*, July 1994, pp. 47–50; "Greenwash, Inc.," *Environmental Action*, Summer 1994, pp. 8–9; and James Ridgeway, "Greenwashing Earth Day," *Village Voice*, April 25, 1995, pp. 15–16.

74. Art Kleiner, "The Greening of Jay Hair," *Garbage*, January/February 1991, p. 3.

75. Carla Koehl and Marc Peyser, "Green Revolt," *Newsweek*, July 10, 1995, p. 6.

76. "NWF: CEO Resigns Amid Alleged 'Internal Conflicts'," *Greenwire*, July 6, 1995, p. 1.

77. Margaret Kriz, "Mark Van Putten: Wildlife Lobby Returns to Its Roots," *National Journal*, January 25, 1997, p. 184.

78. Kriz, "Mark Van Putten: Wildlife Lobby Returns to Its Roots," p. 184.

79. Marylou Tousignant, "Wildlife Federation Neighbors Wary of Sale," *Washington Post*, February 12, 1998, (Virginia Weekly), p. V1.

80. Interview, Frederic D. Krupp, executive director, Environmental Defense Fund, December 23, 1985, New York City, New York, Environmental Defense Fund National Headquarters.

81. Interview, Frederic Krupp.

82. Interview, David Challinor, assistant secretary for science, Smithsonian Institution; board of trustees, Environmental Defense Fund, October 24, 1985, Washington, D.C., Castle Building, Smithsonian.

83. Ronald G. Shaiko, "The Public Interest Dilemma: Organizational Maintenance and Political Representation in the Public Interest Sector" (Ph.D. diss., Syracuse University, 1989), p. 98.

84. Interview, Brian A. Day, director of communications, Environmental Defense Fund, Washington Office, October 15, 1985.

85. Thomas M. Power and Paul Rauber, "The Price of Everything," *Sierra*, November/December 1993, p. 88.

86. "EDF Challenges 'Win-Lose' Thinking," *BNA Environmental Law Update*, July 20, 1992, p. 3.

87. For an excellent analysis of the passage of the 1990 Clean Air Act amendments, including the role of EDF and other organized interests, see Richard E. Cohen, *Washington at Work: Back Rooms and Clean Air* (New York: Macmillan, 1992).

88. Ann Devroy, "Environmental Expert Steals Show at Deregulation Party," *Washington Post*, April 30, 1992, p. A24.

89. "EDF, AD Council, McDonald's Team Up to Encourage Purchase of Recycled Products," *BNA National Environment Daily*, January 12, 1995, p. 1–5.

90. "Dialogue II: Papers Call Deal 'Unusual'; Enviros Cautious," *Greenwire*, July 9, 1992, p. 1.

91. "New EDF Project Will Promote Private-Sector Innovations," *EDF Letter*, March 1996, p. 2.

92. John J. Fried, "Major Environmental Groups Losing Members, Money," New Orleans *Times-Picayune*, December 25, 1994, p. A3.

93. Barbara Ruben and David Lapp, "On the Road from Earth Day: Environmental Action Looks Toward Its Roots to Shape the Future," *Environmental Action*, June 22, 1993, p. 14.

94. Interview, Ruth Caplan, former executive director, Environmental Action, October 22, 1985, Washington, D.C., Environmental Action Headquarters.

95. Margaret Morgan-Hubbard, "Small is Beautiful," *Environmental Action*, June 22, 1993, p. 16. Similar arguments are made by the leaders of women's organizations; see e.g., Anne N. Costain, "Representing Women: The Transition from Social Movement to Interest Group," in Ellen Boneparth, ed., *Women, Power, and Policy*, pp. 19–37 (Elmsford, N.Y.: Pergamon Press, 1982).

96. The "Dirty Dozen" targeting efforts of Environmental Action were far more effective than its PAC expenditures. ENACTPAC never contributed more than $25,000 in any election cycle.

97. "A 501(c)(3) organization cannot endorse, contribute to, work for, or otherwise support a candidate for public office, nor can it oppose one"; see Bob Smucker, *The Nonprofit Lobbying Guide:* Advocating Your Cause—and Getting Results (San Francisco: Jossey-Bass, 1991), p. 85.

98. Interview, Rose Audette, editor, *Environmental Action* magazine, October 21, 1985, Washington, D.C., Environmental Action National Headquarters.

99. Interview, Rose Audette.

100. Interview, Ruth Caplan.

101. Margaret Morgan-Hubbard, "Small Is Beautiful," *Environmental Action*, June 22, 1993, p. 16.

102. Naftali Bendavid, "Environmental Inaction: Death of an Environmental Group," *Legal Times*, December 2, 1996, p. 1.

103. Quoted in Naftali Bendavid, "Environmental Inaction: Death of an Environmental Group," *Legal Times*, December 2, 1996, p. 1.

104. Christopher Boerner and Jennifer Chilton Kallery, "Restructuring Environmental Big Business," ("St. Louis: Center for the Study of American Business,

Washington University, December 1994); see also Barry Shanoff, "Environmental Groups Suffer Fate of Big Business," *World Wastes* (March 1995): 20.

105. Keith Schneider, "Big Environment Hits a Recession," *New York Times*, January 1, 1995, p. 3, col. 4.

106. "Enviro Groups: Wash. U. Study Stirs Debate in Community," *Greenwire*, January 4, 1995, p. 1.

107. Barbara Dudley, "It's Not Easy Being Green," *Mother Jones*, May/June 1995, p. 8.

108. Quoted in Brad Knickerbocker, "Nation's Green Advocates See Their Groups on Critical List," *The Christian Science Monitor,* October 17, 1994, p. 1.

109. Karyn Strickler, "Environmental Towers Build Too High to Keep Grass Roots," *The Christian Science Monitor,* April 21, 1995, p. 18.

110. For a representative sample of the coalitional activities, see Robert Cahn, ed., *An Environmental Agenda for the Future,* (Washington, D.C.: Island Press, 1985), a publication produced by the Group of Ten; Rochelle L. Stanfield, "Environmental Lobby's Changing of the Guard is a Part of Movement's Evolution," *National Journal,* June 8, 1985, pp. 1350–1353; Stanfield, "The Green Blueprint," *National Journal,* July 28, 1988, pp. 1735–1737; "Environmental Organizations Join Coalition to Oppose Federal Subsidies," *BNA National Environment Daily,* March 25, 1993, pp. 1–2; "Environmental Groups Launch Campaign to Counter Threats to Pending Legislation," *BNA National Environment Daily,* July 8, 1994, pp. 1–4; Gary Lee, "Environmentalists Try to Regroup," *Washington Post,* April 22, 1995, p. A3.; Margaret Kriz, "Getting Ready for Round Two, *National Journal,* March 11, 1995, p. 644; "Authors of 'Green Scissors' Report Findings at Briefing," *BNA National Environment Daily,* March 20, 1995, pp. 1–2. National environmental organizations do not always join forces on major policy issues. For example, the debates over the North American Free Trade Agreement (NAFTA) and the General Agreement on Tariffs and Trade (GATT) split the ranks of the national environmental movement.

111. Mark Dowie, *Losing Ground: American Environmentalism at the Close of the Twentieth Century* (Cambridge, Mass.: MIT Press, 1995); for a synopsis of Dowie's argument, see Dowie, "The Fourth Wave: Environmental Movement's Evolution," *Mother Jones,* March 1995, pp. 34–38.

112. Barbara Dudley, "It's Not Easy Being Green," *Mother Jones,* May/June 1995, p. 8. See also Christopher J. Bosso, "Seizing Back the Day: The Challenge of Environmental Activism in the 1990s," in Norman J. Vig and Michael E. Kraft, eds., *Environmental Policy in the 1990s,* 3rd ed., pp. 53–74 (Washington, D.C.: Congressional Quarterly Press, 1997).

4. *Membership Recruitment and Retention: Direct Mail, Telemarketing, and Canvassing*

1. Membership contributions account for more than half of all revenues generated by contemporary national environmental groups; see table 2.6. Christopher J. Bosso, citing data collected by the Conservation Foundation for 248 environmental

groups, finds that 51 percent of revenues of these organizations are generated from membership dues (32 percent) and individual contributions (19 percent). In his own analysis of fourteen national environmental groups, he finds that well over half of the revenues are derived from individuals; see Bosso, "The Color of Money: Environmental Groups and the Pathologies of Fund Raising," in Allan J. Cigler and Burdett A. Loomis, eds., *Interest Group Politics*, 4th ed., pp. 107–108 (Washington, D.C.: Congressional Quarterly Press, 1995).

2. The use of the internet by a growing number of public interest organizations will be analyzed in chapter 6 in the context of membership education and mobilization. There is a recruitment aspect to the worldwide web sites of national organizations, but these home pages are passive recruitment devices, while direct-mail marketing, telemarketing, and canvassing are more direct outreach methods.

3. Alison Cosmedy, ed., *DMA 1994/95 Statistical Fact Book* (Washington, D.C.: Direct Marketing Association, 1994); and Cosmedy, ed., *DMA 1995/96 Statistical Fact Book* (Washington, D.C.: Direct Marketing Association, 1995).

4. Alan Bisbort, "Can't See the Forest For the Junk Mail," *Washington Post*, March 7, 1993, p. C5.

5. Cosmedy, ed., *DMA 1995/96 Statistical Fact Book*, p. 65.

6. Bosso, "The Color of Money: Environmental Groups and the Pathologies of Fund Raising," p. 113.

7. Timothy B. Clark, "After a Decade of Doing Battle, Public Interest Groups Show Their Age," *National Journal*, July 12, 1980, pp. 1136–1141.

8. Richard A. Viguerie, *The New Right: We're Ready to Lead* (Falls Church, Va.: Viguerie, 1980), p. 120.

9. R. Kenneth Godwin, *One Billion Dollars of Influence: The Direct Marketing of Politics* (Chatham, N.J.: Chatham House, 1988).

10. Roger Craver, quoted in Susan Rouder, "Mobilization by Mail," *Citizen Participation*, September/October 1980, p. 4.

11. See, e.g., Kim Klein, "Twenty Words that Sell," *Grassroots Fundraising Journal* 6 (February 1987): 13–14; Jean Harris, "High-Power Words," *Direct Marketing* 51 (1989): 51, 79–80; Richard G. Ensman Jr., "The Art and Science of Direct Mail Copywriting: Part One," *Grassroots Fundraising Journal* 6 (April 1987): 3–7.

12. Environmental Action did not offer an affinity credit card. This premium, offered by an increasing number of public interest organizations, as well as political parties, generates revenues for the organization, based on a small (1–2) percentage of service charges on each purchase made by the cardholder being contributed to the organization by the financial institution managing the card. A recent USPS ruling, however, has had a chilling effect on the use of affinity cards. "USPS announced that the use of a word identifying an association's affinity credit card program, such as Visa or Mastercard, destroys nonprofit mail eligibility." See "Issues Update: Nonprofit Mail," *Association Management* 50 (1) (January 1998): 9.

13. National Wildlife Federation, Document #H1501.

14. The National Committee for the Preservation of Social Security and

Medicare also mails similarly large pieces of mail, with correspondingly large printing to suit their aged constituents' needs.

15. The process of list rental is guided by established rules regarding the use of rented names and addresses. The most common agreement reached between the renter and the rentee limits the use of the list to a single mailing, with the mail pieces dropped on an agreed upon date, with payment for the use of the list due thirty days following the mailing. It is often the case that volume or "net name" discounts are offered when the renter seeks hundreds of thousands of names and addresses.

16. Interview, Vicky Monrean, membership development, Greenpeace, U.S.A, Washington, D.C., June 29, 1988; see also Ronald G. Shaiko, "Greenpeace U.S.A.: Something Old, New, Borrowed," *The Annals* 528 (Summer 1993): 93.

17. Telephone interview, Richard Hammond, president, Names in the News, San Francisco, California, April 6, 1986.

18. Julian Keniry, "Environmental Movement Booming on Campus," *Change*, September/October 1993, p. 42.

19. Keniry, "Environmental Movement Booming on Campus," pp. 44–48; Alice Dembner, "Movement is Strong on Campus," *Boston Globe*, November 12, 1994, p. 28; and Thomas M. Parris, "A Plethora of Campus Environmental Initiatives on the Net," *Environment* 40 (7) (September 1998): 3.

20. Lawrence S. Rothenberg, in his analysis of the membership distribution at Common Cause, finds similar patterns: "The propensity of the association's contributors to come from urban areas is evidenced by a pronounced concentration of members in the more urban Northeast and Far West (Figure 3.1). Although the likelihood that a member of the voting-age population in these areas belongs to Common Cause is about 1 in 100, the probability in other areas shrinks to roughly 1 in 500." See Rothenberg, *Linking Citizens to Government: Interest Group Politics at Common Cause* (New York: Cambridge University Press, 1992), pp. 31–35.

21. Frank Clifford, "Environmental Movement Struggling As Clout Fades," *Los Angeles Times*, September 21, 1994, p. A1.

22. Prior to March of 1996, the prohibition of electioneering by 501(c)(3) organizations had not been applied to direct-mail solicitations conducted by such organizations.

23. Damon Chappie, "The IRS's 'Story of M' May Affect '96 Politics," *Roll Call*, April 15, 1996, pp. 1, 31.

24. Chappie, "The IRS's 'Story of M' May Affect '96 Politics," p. 31.

25. Chappie, p. 31.

26. Interview, Alden Meyer, executive director, League of Conservation Voters, March 28, 1986, Washington, D.C., League of Conservation Voters National Headquarters.

27. Fred Krupp of the Environmental Defense offered his membership some guidance regarding the overwhelming nature of direct mail that deserves repeating to a wider audience: "If you prefer not to receive commercial catalogs and advertis-

ing mail, you can remove your name from most lists used to recruit new members. Just send your name and address (and any variations) to: Mail Preference Service, c/o DMA, P.O. Box 9008, Farmingdale, NY 11735–9008." EDF and most organizations also provide "No Exchange" boxes on their membership forms that remove the names of individuals who do not want to be solicited as a result of list rentals; see Fred Krupp, "Director's Message: Please, Mr. Postman," *EDF Letter,* March 1996, p. 3.

28. Correspondence, American Telemarketing Association, Los Angeles, California, March 1996.

29. Daniel Borochoff, "The Latest Patterns of Telemarketing Abuse," *AIP Charity Rating Guide and Watchdog Report,* Winter 1995, p. 1.

30. Borochoff, "The Latest Patterns of Telemarketing Abuse," p. 1.

31. Interview, Kathy Swayze, partner, Herzog Swayze, Inc., Washington, D.C., March 8, 1996.

32. In 1997 I received more than fifty telemarketing appeals from Environmental Defense Fund, The Wilderness Society, and the Sierra Club as well as from a host of other public interest, political, police or fire, and service associations, and various colleges and universities. This does not include the numerous solicitations from MCI, Sprint, and AT&T.

33. See Kenneth Godwin and Robert C. Mitchell, "The Impact of Direct Mail on Political Organizations," *Social Science Quarterly* 65 (1984): 829–839.

34. Sheldon Wolin, quoted in Michael W. McCann, *Taking Reform Seriously* (Ithaca, N.Y.: Cornell University Press, 1986), p. 300.

35. Marvin A. Jolson, *Consumer Attitudes Toward Direct-to-Home Marketing Systems* (New York: Dunellen, 1970); and Steve Cannizaro, "Con Men Target Elderly in Roof Repair Scam," New Orleans *Times-Picayune,* January 25, 1996, p. B1.

36. See, e.g., Stephen Butterfield, *Amway: The Cult of Free Enterprise* (Boston: South End Press, 1985).

37. Roughly three-quarters of these salespersons are women; virtually all are independent contractors rather than employees of the corporations that manufacture the products. More than 90 percent of these individuals are considered full-time workers. Data from the Direct Selling Association made available by Liz Doherty, director of communications, Direct Selling Association, Washington, D.C., March 1996.

38. Quoted in B. J. Bergman, "Reaching Out, One by One," *Sierra,* January/February 1997, p. 53.

39. See, e.g., Jeffrey M. Berry, *Lobbying for the People: The Political Behavior of Public Interest Groups* (Princeton, N.J.: Princeton University Press, 1977), pp. 18–78; Jack L. Walker, Jr., *Mobilizing Interest Groups in America* (Ann Arbor: University of Michigan Press, 1991), pp. 75–103; Allan J. Cigler and Anthony J. Nownes, "Public Interest Entrepreneurs and Group Patrons," in Allan J. Cigler and Burdett A. Loomis, *Interest Group Politics,* 4th ed., pp. 77–100 (Washington, D.C.: Congres-

sional Quarterly Press, 1995); Bosso, "The Color of Money: Environmental Groups and the Pathologies of Fund Raising," pp. 117–122.

40. Berry, *Lobbying for the People*, pp. 71–76; Walker, *Mobilizing Interest Groups in America*, pp. 77–81.

41. Paul Starobin, "Raging Moderates," *National Journal*, May 10, 1997, pp. 914–918.

42. "1998 Earth Share Member Organizations," *Earth Share*, Washington, D.C., fall 1998, pp. 1–7.

43. John C. Stauber and Sheldon Rampton, "Green PR: Silencing Spring," *Environmental Action* (Winter 1996): 16–19; see also Bosso, "The Color of Money," pp. 119–122; Margaret Morgan-Hubbard, "Money and Environmental Groups: How Clean is 'Green?'" *Environmental Action* (Winter 1996): 20–21; and Morgan-Hubbard, "Green Backers," *Environmental Action* (Winter 1996): 22–23. In the last article, the corporate contributions to eleven environmental organizations, including the Sierra Club and the National Wildlife Federation, are analyzed.

44. "Sierra Club Financial Report," *Sierra*, September/October 1995, pp. 102–103.

45. Berry, *Lobbying for the People*, pp. 45–59; Douglas W. Costain and Ann N. Costain, "Representing Women: The Transition from Social Movement to Interest Group," *Western Political Quarterly* 34 (1981): 100–113; Richard Scotch, *From Goodwill to Civil Rights: Transforming Federal Disability Policy* (Philadelphia: Temple University Press, 1985); Georgia Duerst-Lahti, "The Government's Role in Building the Women's Movement," *Political Science Quarterly* 104 (1989): 249–268.

46. Walker, *Mobilizing Interest Groups in America*, p. 82.

47. Clifford W. Brown Jr., Lynda W. Powell, and Clyde Wilcox identify wealthy supporters of presidential candidates in the nomination process as contributors of serious money. In this context they define such donors as individuals giving from $200 to the legal limit of $1,000 in each campaign cycle. For our purposes, serious money contributors are individuals giving at least $1,000 annually to an organization and more likely giving several thousand dollars. See Brown Jr., Powell, and Wilcox, *Serious Money: Fundraising and Contributing in Presidential Nominating Campaigns* (New York: Cambridge University Press, 1995), p. 8.

48. A growing number of organizations are developing direct video campaigns to be used in small group settings to solicit new members and contributors through existing social networks such as Rotary, Kiwanis, Jaycees, and other service and community organizations. Environmental Defense Fund, for example, produced a "25-minute video newsletter" to celebrate its twentieth anniversary: "Environmental Vision: Twenty Years of Innovative Solutions." The video presentation was hosted by John Chancellor and narrated by Sally Field and was funded by grants from the Rockefeller Family Fund and the Benton Foundation.

49. Brown Jr., Powell, and Wilcox, *Serious Money*, p. 64; emphasis in original text.

50. I have found that the Office of Charities Registration for the State of New York consistently has the most comprehensive information system available for public access. The Office of Charities Registration is located at 162 Washington Avenue, Albany, NY 12231.

51. Ronald G. Shaiko, "The Public Interest Dilemma: Organizational Maintenance and Political Representation in the Public Interest Sector" (Ph.D. diss., Syracuse University, 1989), p. 239.

52. For more information on AIP, contact the American Institute of Philanthropy at 4579 Laclede Avenue, Suite 136, St. Louis, Missouri 63108.

53. "Getting the Most from Your Rating Guide," *AIP Charity Rating Guide and Watchdog Report,* Summer/Fall 1994, pp. 3–4.

54. Boys Town (Father Flanagan's Boys' Homes) has filed a lawsuit against AIP due to its rating of "F" by AIP. AIP based its rating of Boys Town on the fact that the organization has amassed over $580 million in assets, enough for the entire operation to exist for almost six years. By AIP standards, Boys Town should be judged as "least needy" among youth-residential care charities. Borochoff argues that "Boys Town's lawsuit will not deter AIP from providing the information that you need to make intelligent giving decisions." See Daniel Borochoff, "Think! . . . Before You Give," *AIP Charity Rating Guide and Watchdog Report,* Summer/Fall 1994, pp. 1–2.

5. It's Not Easy Being Green: Leadership Incentives and Membership Motivations

1. James Q. Wilson, *Political Organizations* (Princeton, N.J.: Princeton University Press, 1995), pp. vii–xxiv. Wilson provided the first systematic incentive theory that has been incorporated into virtually all subsequent research on the topic. His dimensions include material, solidary, and purposive incentives; see Peter B. Clark and James Q. Wilson, "Incentive Systems: A Theory of Organizations," *Administrative Science Quarterly* 6 (1961): 129–166.

2. See Mancur Olson, *The Logic of Collective Action: Public Goods and the Theory of Groups* (Cambridge, Mass.: Harvard University Press, 1965).

3. Terry M. Moe, *The Organization of Interests: Incentives and the Internal Dynamics of Political Interest Groups* (Chicago: University of Chicago Press, 1980); and Moe, "A Calculus of Group Membership," *American Journal of Political Science* 24 (1980): 593–631.

4. See David Marsh, "On Joining Interest Groups," *British Journal of Political Science* 6 (1976): 257–271; and William P. Browne, "Benefits and Membership: A Reappraisal of Interest Group Activity," *Western Political Quarterly* 29 (1976): 258–273; see also David King and Jack L. Walker, "The Provision of Benefits by Interest Groups in the United States," *Journal of Politics* 54 (1992): 394–426.

5. The term *free riding* is associated with Olson's work but never appears in *The Logic of Collective Action.* Olson does discuss the phenomenon, though, without specifically labeling the action: "Once a smaller member has the amount of the collective good he gets free from the largest member, he has more than he would have

purchased for himself, and has no incentive to obtain any of the collective good at his own expense" (p. 35).

6. For additional research on noneconomic incentives from a sociological perspective, see David Knoke, *Organizing for Collective Action: The Political Economies of Associations* (New York: Aldine de Gruyter, 1990); David Knoke and James R. Wood, *Organized for Action: Commitment in Voluntary Associations* (New Brunswick, N.J.: Rutgers University Press, 1981); and Knoke and Christine Wright-Isak, "Individual Motives and Organizational Incentive Systems," in Samuel B. Bacharach, ed., *Research in the Sociology of Organizations*, vol. 1, pp. 209–253 (Greenwich, Conn.: JAI Press, 1981).

7. Lawrence S. Rothenberg, "Organizational Maintenance and the Retention Decision in Groups," *American Political Science Review* 82 (1988): 1129–1152; and Rothenberg, *Linking Citizens to Government: Interest Group Politics at Common Cause* (New York: Cambridge University Press, 1992).

8. John Mark Hansen, "The Political Economy of Group Membership," *American Political Science Review* 79 (1985): 79–86.

9. The groups and the time frame for the analysis were determined by the collection of membership data by sociologist Robert C. Mitchell for Resources for the Future. In April 1978 Mitchell sent mail surveys to a random sample of 1,000 members in each of the five groups. The survey produced a response of 3,128, for a return rate of 62.6% across the groups. By group, the total number of responses are as follows: Environmental Action (EA), N = 705; Environmental Defense Fund (EDF), N = 630; Wildlife Federation (NWF), N = 509; Sierra Club (SC), N = 661; and The Wilderness Society (TWS), N = 623. The data set was obtained through the Roper Organization, University of Connecticut, Storrs: Roper Archive File: USMISCRFFENVRN78.

10. See R. Kenneth Godwin and Robert C. Mitchell, "Rational Models, Collective Goods, and Nonelectoral Political Behavior," *Western Political Quarterly* 35 (1982): 161–180; and V. Kerry Smith, "A Theoretical Analysis of the 'Green Lobby'," *American Political Science Review* 79 (1985): 132–147. Both analyses aggregate these resulting data sets drawn from the five groups, assuming that all the respondents are similar in attitudes and motivations.

11. See Talcott Parsons, *The Structure of Social Action* (New York: Free Press, 1937); for a synthesis of Parsons, see Jeffrey C. Alexander, "Formal and Substantive Voluntarism in the Work of Talcott Parsons: A Theoretical and Ideological Reinterpretation," *American Sociological Review* 43 (1978): 177–198.

12. Robert C. Mitchell, "National Environmental Lobbies and the Apparent Illogic of Collective Action," in Clifford Russell, ed., *Collective Decision Making*, pp. 87–121 (Baltimore: Johns Hopkins University Press, 1979).

13. See, e.g., Susan Shott, "Emotion and Social Life: A Symbolic Interactionist Analysis," *American Journal of Sociology* 84 (1979): 1317–1334; Arlie Russell Hochschild, "Emotion Work, Feeling Rules, and Social Structure," *American Journal of Sociology* 85 (1979): 551–574.

14. Knoke and Wright-Isak, "Individual Motives and Organizational Incentive Systems," pp. 228.

15. Cf., Harriet Tillock and Denton E. Morrison, "Group Size and Contributions to Collective Action: A Test of Mancur Olson's Theory of Zero Population Growth, Inc.," in Louis Kriesburg, ed., *Research in Social Movements, Conflict, and Change,* vol. 2, pp. 131–158 (Greenwich, Conn.: JAI Press, 1979); see Olson's letter to the authors in the same volume, pp. 149–150; Gerald Marwell and Ruth E. Ames, "Experiments on the Provision of Public Goods: I. Resources, Interest, Group Size, and the Free Rider Problem," *American Journal of Sociology* 84 (1979): 1334–1360; and Edward J. Walsh and Rex H. Warland, "Social Movement Involvement in the Wake of a Nuclear Accident: Activists and Free Riders in the TMI Area," *American Sociological Review* 48 (1983): 764–780.

16. Jeffrey Berry, *Lobbying for the People: The Political Behavior of Public Interest Groups* (Princeton: Princeton University Press, 1977).

17. Moe, *The Organization of Interests,* chapter 2: "The Decision to Join."

18. The composite characteristics presented above are derived from demographic and attributional data found in the Mitchell data set (see table 5.1).

19. See Smith, "A Theoretical Analysis of the 'Green Lobby,'" pp. 132–147.

20. See appendix B for the operationalization of the variables used in the following empirical analyses. The validity and reliability of these measures are discussed in appendix C.

21. The *THREAT* variable is operationalized using questions that are not group-specific; therefore, it would be inaccurate to link this motivation to an organizationally controlled incentive. Although these organizations, as well as the vast majority of organizations using direct-mail recruiting, stress threats to the environment to motivate potential members to join their groups, it is not possible to link each survey respondent with the specific messages that were used by organization leaders to attract this member.

22. Direct-mail correspondence received from the Environmental Defense Fund. The Environmental Defense Fund has more recently invested more resources in its recruitment and retention efforts.

23. Respondents were asked how appropriate or inappropriate the descriptions "environmentalist" and "conservationist" would be if applied to them personally. Since the vast majority of respondents report that their views on environmental and conservation issues were already developed prior to joining their groups, this self-descriptive measure seems an appropriate gauge of this direction of individuals' purposive inclinations.

24. See "The DISCRIM Procedure," in Alice Allen Ray, ed., *SAS User's Guide: Statistics,* pp. 381–396 (Cary, N.C.: SAS Institute, 1982).

25. This conclusion was based on a variety of interviews with membership development directors as well as with several list brokers.

26. For the models in tables 5.3, 5.4, 5.5, and 5.6, the estimated regression coefficients (maximum likelihood estimates, MLE), estimated standard errors for each

MLE, -2 log likelihood ratio chi-square model statistic, the individual variable chi-square statistics, the percentage of correct predictions, and approximated R are presented. See Frank E. Harrell Jr., "The LOGIST procedure," in *SUGI Supplemental Library User's Guide*, pp. 181–202 (Cary, N.C.: SAS Institute, 1983). Additive linear models are proposed.

27. R. Kenneth Godwin and Robert Cameron Mitchell, "The Impact of Direct Mail for Political Organizations," *Social Science Quarterly* 65 (1984): 829–839. The National Wildlife Federation is now one of the largest public interest direct mailers in the country, sending out over 200 million pieces annually.

28. See Jay D. Hair and Patrick A. Parenteau, "Special Report: The Results of the National Wildlife Federation's Associate Membership Survey and Affiliate Leadership Poll on Major Conservation Policies," (Washington, D.C.: National Wildlife Federation, July 14, 1981).

29. See Robert Salisbury, "An Exchange Theory of Interest Groups," *Midwest Journal of Political Science* 13 (1969): 1–32.

30. See Knoke and Wright-Isak, "Individual Motives and Organizational Incentive Systems," for their hierarchical ranking of individual motivations.

31. Since the measures employed are exogenous (non-group-specific), members falling into the control group (coded o) who also belong to NWF were eliminated from the analysis.

32. For an analysis of the impact of religion on environmental concern, see Ronald G. Shaiko, "Religion, Politics, and Environmental Concern: A Powerful Mix of Passions," *Social Science Quarterly* 68 (1987): 244–262.

33. Again, since variables are exogenous, members in the control group belonging to EDF are eliminated from the analysis.

34. During an eighteen-month period, membership in these five organizations produced more than 150 direct-mail solicitations from more than 65 groups, most of which were environmental groups but also included Mothers Against Drunk Driving, Gray Panthers, American Civil Liberties Union, Common Cause, Amnesty International, Handgun Control, Inc., and The Humane Society.

35. Survey respondents were given a list of thirty-one environmental/conservation organizations. For each group respondents were asked whether or not they recognize the organization and whether or not they belong to the organization. The methodology employed in table 5.7 differs from the earlier analyses in that the dependent variable is simply the number of organizations to which each respondent belonged, rather than the dichotomous (0 or 1) dependent variables. Significant variables are again presented in italics with the most significant variables identified with asterisks.

36. In an earlier effort the author did attempt to address the question of longevity motivation using this cross-sectional data in a model that included length of membership as the dependent variable. Separate models were run for each group utilizing the four motivational incentives. For NWF and EDF members, a significant positive relationship was identified on the goals dimension. That

is, the longer one's tenure, the greater the importance of the organization's policy goals (NWF, p = 0.015; EDF p = 0.026). For EA members, longer term members are motivated by premiums more than members with less tenure in the organization. While these findings do not capture the true dynamic of individual changes in motivations over time, they do provide some evidence of the impact of time on motivational behavior.

37. For methodological critique of the validity and reliability of the preceding analyses, see appendix C.

6. The Heart of the Matter: Leadership-Membership Connections

1. Michael T. Hayes, "Interest Groups: Pluralism or Mass Society," in Allan J. Cigler and Burdett A. Loomis, eds., Interest Group Politics, p. 111 (Washington, D.C.: Congressional Quarterly Press, 1983).

2. John W. Gardner, On Leadership (New York: Free Press, 1990), pp. 23–37.

3. Kay Lehman Schlozman and John T. Tierney, Organized Interests and American Democracy (New York: Harper and Row, 1986), p. 197.

4. Peter H. Odegard, Pressure Politics: The Story of the Anti-Saloon League (New York: Columbia University Press, 1928); see also Pendleton Herring, Group Representation Before Congress (Baltimore: Johns Hopkins University Press, 1929); and Noble E. Cunningham, Circular Letters of Congressmen to their Constituents, 1789–1829 (Chapel Hill: University of North Carolina Press, 1978).

5. Bill Keller, "Special-Interest Lobbies Cultivate the 'Grassroots' to Influence Capitol Hill," Congressional Quarterly Weekly Report, September 12, 1981, p. 1741; Keller, "Computers and Laser Printers Have Recast the Injunction: 'Write Your Congressman'," Congressional Quarterly Weekly Report, September 12, 1982, p. 2247; Keller, "Lowest Common Denominator Lobbying: Why the Banks Fought Withholding," Washington Monthly, May 1983, pp. 32–39; and Deborah Baldwin, "Ideology by Mail," The New Republic, July 7–14, 1979, p. 20. For more recent accounts of orchestrated grassroots lobbying efforts, see Stephen Engelberg, "A New Breed of Hired Hands Cultivates Grass-Roots Anger," New York Times, March 17, 1993, pp. A1, A17; and Jodi Enda, "Organizing the Grass Roots to Plant a Seed in Washington," Philadelphia Inquirer, October 15, 1995, pp. A1, A8.

6. Kay Lehman Schlozman and John T. Tierney, "More of the Same: Washington Pressure Groups in a Decade of Change," The Journal of Politics 45 (1983): 351–377.

7. Ronald G. Shaiko, "Petitioning Government: Constituent Communications and Congressional Responses," presented at the Annual Meeting of the Southern Political Science Association, Charlotte, North Carolina, November 5–7, 1987, p. 12. Cf., John Kingdon, Congressmen's Voting Decisions (New York: Harper and Row, 1973); and Kingdon, Agendas, Alternatives, and Public Policies (Boston: Little, Brown, 1984).

8. Bonner and Associates/Gallup Poll, "Survey of Business Issues in the New Congress" (January 1993), pp. 10–11.

9. Jeffrey M. Berry, *The Interest Group Society* (Boston: Little, Brown, 1984), p. 150; see also Burson-Marsteller, "Communications and Congress: A Study of the Exposure of Federal Legislative Offices to Various Information Vehicles" (Washington, D.C.: Public Affairs/Government Relations, Burson-Marsteller, 1981).

10. Quoted in Burdett A. Loomis, "A New Era: Groups and the Grass Roots," in Allan J. Cigler and Burdett A. Loomis, eds, *Interest Group Politics*, p. 180 (Washington, D.C.: Congressional Quarterly Press, 1983). This argument is explored further in Linda L. Fowler and Ronald G. Shaiko, "The Grass Roots Connection: Environmental Activists and Senate Roll Calls," *American Journal of Political Science* 31 (1987): 484–510.

11. R. Kenneth Godwin, "Lobbying Choices of Citizen Action Groups," unpublished manuscript, 1984, p. 6.

12. Andrew S. McFarland, *Common Cause: Lobbying in the Public Interest* (Chatham, N.J.: Chatham House, 1984), pp. 64–68. See also, David Knoke, *Organizing for Collective Action: The Political Economies of Associations* (New York: Aldine de Gruyter, 1990), pp. 187–213.

13. Quoted in Loomis, "A New Era: Groups and the Grass Roots," p. 184.

14. Cf., Jack L. Walker, *Mobilizing Interest Groups in America*, (Ann Arbor: University of Michigan Press, 1991), p. 49. Walker distinguishes between "autonomous individuals" and "organizational representatives."

15. Compare Charles O. Jones, *Clean Air: The Policies and Politics of Pollution Control* (Pittsburgh: University of Pittsburgh Press, 1975); Bernard Asbell, *The Senate Nobody Knows* (Garden City, N.Y.: Doubleday, 1978); and Norman J. Ornstein and Shirley Elder, *Interest Groups, Lobbying, and Policymaking* (Washington, D.C.: Congressional Quarterly Press, 1978). Also see Christopher J. Bosso, *Pesticides and Politics: The Life Cycle of a Public Issue* (Pittsburgh: University of Pittsburgh Press, 1987). For an assessment of the changes in the relationship between environmentalists and government in the case of the 1990 Clean Air Act, see Richard E. Cohen, *Washington at Work: Back Rooms and Clean Air* (New York: MacMillan, 1992).

16. Riley E. Dunlap and Michael Allen, "Partisan Differences on Environmental Issues: A Congressional Roll Call Analysis," *Western Political Quarterly* 29 (1976): 384–397; John E. Jackson, *Constituents and Leaders in Congress* (Cambridge: Harvard University Press, 1974); and Lawrence S. Rothenberg, *Linking Citizens to Government: Interest Group Politics at Common Cause* (New York: Cambridge University Press, 1992).

17. Jeffrey M. Berry, *Lobbying for the People: The Political Behavior of Interest Groups*. (Princeton: Princeton University Press, 1977), pp. 43–78.

18. Kay Lehman Schlozman and John T. Tierney, "More of the Same," p. 357. In fact, since 1990 I have conducted more than two hundred interviews with Washington lobbyists representing every sector of the interest group network and have found that coalition formation and grassroots lobbying are the two strategies most often mentioned by representatives of economic interests and those representing public interest organizations.

19. Raymond A. Bauer, Ithiel de Sola Pool, and Lewis A. Dexter, *American Business and Public Policy* (New York: Atherton Press, 1963).

20. James N. Rosenau, *Citizenship Between Elections* (New York: Prentice-Hall, 1973).

21. James Q. Wilson, *Political Organizations* (New York: Basic Books, 1973).

22. See e.g., John R. Wright, *Interest Groups and Congress: Lobbying, Contributions, and Influence* (Boston: Allyn and Bacon, 1996).

23. Loomis, "A New Era: Groups and the Grass Roots."

24. Godwin, "Lobbying Choices of Citizen Action Groups"; McFarland, *Common Cause: Lobbying in the Public Interest;* and Rothenberg, *Linking Citizens to Government: Interest Group Politics at Common Cause.*

25. Cf., Charles E. Clapp, *The Congressman: His Work as He Sees It* (Washington, D.C.: Brookings Institution, 1963); Burson-Marsteller, "Communications and Congress: A Study of the Exposure of Federal Legislative Offices to Various Information Vehicles"; and Bonner and Associates/Gallup Poll, "Survey of Business Issues in the New Congress."

26. Andrew S. McFarland, *Common Cause,* pp. 64–68.

27. Richard F. Fenno Jr., *Homestyle: House Members in their Districts* (Boston: Little, Brown, 1978).

28. James B. Kau and Paul H. Rubin, "Public Interest Lobbies: Membership and Influence," *Public Choice* 34 (1979): 45–54.

29. Cf., John E. Jackson, *Constituents and Leaders in Congress;* Gregory B. Marcus, "Electoral Coalitions and Senate Roll Call Behavior: An Ecological Analysis," *American Journal of Political Science* 18 (1974): 595–607; also Robert Crandall, "The Use of Environmental Policy to Reduce Economic Growth in the Sunbelt: The Role of Electric Utility Rates," in Michael Crew, ed., *Regulatory Reform and Public Utilities,* 125–140 (Lexington, Mass.: Lexington Books, 1982); and Sam Peltzman, "Constituent Interest and Congressional Voting," *The Journal of Law & Economics* 27 (April 1984): 181–210.

30. Terry M. Moe, *The Organization of Interests: Incentives and the Internal Dynamics of Political Interest Groups* (Chicago: University of Chicago Press, 1980), pp. 206–213.

31. Fowler and Shaiko, "The Grass Roots Connection: Environmental Activists and Senate Roll Calls."

32. R. Kenneth Godwin, *One Billion Dollars of Influence* (Chatham, N.J.: Chatham House, 1988), pp. 43–72.

33. Telephone Interview, Richard Hammond, president, Names in the News, San Francisco, California, April 6, 1986.

34. Mitchell, "National Environmental Lobbies and the Apparent Illogic of Collective Action," In Clifford Russell, ed., *Collective Decision Making,* p. 109 (Baltimore: Johns Hopkins University Press, 1979).

35. Interview, Joel Thomas, general counsel, National Wildlife Federation, March 8, 1986, Vienna, Virginia, National Wildlife Federation Headquarters.

36. See Ronald G. Shaiko, "Interest Group Activity at the Grass Roots Level: A United Front?" presented at the Annual Meeting of the Southern Political Science Association, Savannah, Georgia, November 1–3, 1984.

37. See Bill Gifford, "Merchandise the Planet: The National Wildlife Federation Profits by Nature," *Washington City Paper*, April 24, 1992, pp. 10, 12, for a critique of the "educational" merchandise sold by NWF.

38. Thus far, the coding scheme has been fairly straightforward, the coding categories being self-explanatory. However, a distinction must be made between general policy concerns and specific environmental policy issues. This distinction is quite clear in virtually all cases. The distinction is based on the presence or absence of a specific topic under debate in a specific institutional arena. For example, a note or article on toxic waste or endangered species is considered as a general policy concern. The reader is presented with information about an environmental issue, but with no specific mention of legislation, executive branch activity, or litigation. Conversely, an article on the Endangered American Wilderness Act, amendment to the Endangered Species Act, or the Carter Energy Package, is considered a specific policy issue.

39. Due to a shift in publication months, ten issues were analyzed for content, capturing a thirteen-month period in 1977–1978.

40. The results reported in tables 6.2 and 6.3 for The Wilderness Society include the combined analyses of *Living Wilderness* and *Wilderness Report* magazines.

41. See appendix D for an analysis of the specific and general policy issues presented in the 1977–1978 communications of the five organizations.

42. Shaiko, "Interest Group Activity at the Grass Roots Level: A United Front?"

43. Sample respondents were given a list of twenty-two conservation/environmental issues and approaches. They were asked to check each of the items that is "very important to you personally." At the end of the list they were asked to place the number of the issue response that is "most important to you personally."

44. Population problems were not discussed in the magazines during the year of the content analysis; however, the issue appears on many membership lists as a high priority. With the average length of membership being about three years, many members joined or were affiliated during an earlier period in which there was discussion of this issue in the communications. This may indicate that there is some degree of consistency in the members' issue agendas over time.

45. See, e.g., Sidney Verba, Kay Lehman Schlozman, and Henry E. Brady, *Voice and Equality: Civic Voluntarism in American Politics* (Cambridge, Mass.: Harvard University Press, 1995), pp. 186–266.

46. See appendix E for operationalization of issues presented in table 6.6.

47. Seven issues were included in the 1977–1978 analysis due to the timing of the Mitchell survey. These seven magazines covered a thirteen-month period.

48. Interview, Robert Strohm, editor in chief, *National Wildlife*, October 27, 1992.

49. *Wilderness* was published quarterly in 1991–1992.

50. In 1996 EDF published six issues of *EDF Letter*.

51. Interview, Robert Strohm, editor in chief, *National Wildlife*.

52. These tensions are also manifest within other national environmental organizations. In early 1996 the new president of the Audubon Society, John Flicker, pulled an article, written by *New York Times* columnist Tom Wicker, on the environmental record of President Clinton. The article had already been accepted for publication in *Audubon* and paid for by editor Michael Robbins; see Howard Kurtz, "Audubon Chief Pulls Article on Clinton," *Washington Post,* February 10, 1996, p. C7.

53. At the fundamental level, the internet is a global network of computers. The internet has come to mean all activity and interaction that takes place on the network, such as e-mail and the worldwide web (WWW). The WWW is an Internet service that has grown to dominate information distribution on the internet; it is composed of thousands of individual websites connected to one another via hyperlinks. A website is simply a location on the WWW dedicated to a specific purpose; see "Glossary: Politics on Line Conference," April 17, 1996, Omni Shoreham Hotel, Washington, D.C.

54. Some experts predict that there will be one billion internet users worldwide by the year 2005. See Robert O. Keohane and Joseph S. Nye Jr., "Power and Interdependence in the Information Age," *Foreign Affairs* 77 (5) (September/October 1998): 82. See also Kara Swisher, "There's No Place Like a Home Page," *Washington Post,* July 1, 1996, pp. A1, A8; Robert Kuttner, "Wired for Democracy," *Boston Globe*, February 10, 1996, p. 15; and Graeme Browning, *Electronic Democracy: Using the Internet in American Politics* (Hartford, Conn.: Online, 1996).

55. Interview, Glen Caroline, director, Grassroots Programs, National Rifle Association, Institute for Legislative Action, Washington, D.C., May 23, 1996.

7. Organizational Leadership and Grassroots Empowerment: Reinvigorating Public Interest Representation in the United States

1. Some organizations send as many as ten renewal notices before giving up on a membership. Other organizations incorporate telemarketing as a means of increasing renewal rates. The Sierra Club has raised its retention rate by 2 percent through telemarketing.

2. Peter Steinhart, "What Can We Do About Environmental Racism?" *Audubon* 93 (4) (May 1991): 18–20; and Catalina Camisa, "Poor, Minorities Want Voice in Environmental Choices," *Congressional Quarterly Weekly Report,* August 21, 1993, pp. 2257–2260.

3. See, e.g., Jonathan Collett and Stephen Karakashian, *Greening the College Curriculum: A Guide to Environmental Teaching in the Liberal Arts* (Washington, D.C.: Island Press, 1996).

4. See Richard Himelfarb, *Catastrophic Politics: The Rise and Fall of the Medicare*

Catastrophic Coverage Act of 1988 (University Park: Pennsylvania State University Press, 1995).

5. "Opinions of Americans Age 45 and Over on the Medicare Catastrophic Coverage Act" (Washington, D.C.: Hamilton, Frederick & Schneiders for AARP/AARP Research & Data Resources Department, January 1989).

6. Focus group research entails moderating a discussion of a small group, usually including eight to twelve individuals (e.g., members of an organization). Rather than asking direct questions in a survey format, the moderator conducts a discussion lasting an hour or two and based on a predetermined issue agenda.

While focus groups do not provide statistically valid and reliable data applicable to the larger membership, they do provide richer and more detailed evaluations of issues. Focus groups are often used to test commercials used in issue campaigns conducted by political organizations and to test changes in magazine content, format, and presentation as well as issues to be used in direct-mail campaigns. See, e.g., Michael W. Traugott and Paul J. Lavrakas, *The Voter's Guide to Election Polls* (Chatham, N.J.: Chatham House, 1996), pp. 34–35.

7. Interview, Thomas Halatyn, director of marketing, Campaign Mail & Data, Inc., Washington, D.C., May 16, 1996.

8. John R. Harbison and Peter Pekar Jr., *A Practical Guide for Leapfrogging the Learning Curve* (Los Angeles: Booz-Allen and Hamilton, 1993).

9. George R. Milne, Easwar S. Iyer, and Sara Gooding-Williams, "Environmental Organization Alliance Relationships Within and Across Nonprofit, Business, and Government Sectors," *Journal of Public Policy and Marketing* (Fall 1996): 375–403.

10. Rue E. Gordon, ed., *1993 Conservation Directory* (Washington, D.C.: National Wildlife Federation, 1993).

11. Milne, Easwar, and Gooding-Williams, "Environmental Organization Alliance Relationships Within and Across Nonprofit, Business, and Government Sectors," p. 377.

12. Scott Allen, "The Greening of the Movement: Big Money is Bankrolling Select Environmental Causes," *Boston Globe*, October 19, 1997, p. A1.

13. Allen, "The Greening of the Movement."

14. Quoted in Allen, "The Greening of the Movement."

15. For analysis of the Sierra Club debate and national membership vote on immigration policy, see William Branigin, "Sierra Club Votes for Neutrality on Immigration: Population Issue 'Intensely Debated,'" *Washington Post*, April 26, 1998, p. A16; and Brad Knickerbocker, "Environment vs. Immigrants," *The Christian Science Monitor*, April 27, 1998, p. 3.

16. Gina Neff, "Eco-Efficiency: The Business Link to Sustainable Development, *The Nation*, November 7, 1997, p. 50.

17. James Q. Wilson, *Political Organizations* (Princeton, N.J.: Princeton University Press, 1995), p. x.

18. See, e.g., Russell Hardin, *Collective Action* (Baltimore, Md.: Johns Hopkins University Press, 1982), pp. 101–124.

19. Information presented on Common Cause was derived from informal conversations with my colleague, Candice Nelson, member of the national governing board of Common Cause and from an interview with Donald Simon, executive vice president, Common Cause, Washington, D.C., June 25, 1996.

Appendixes C and D

1. See, e.g., Jarol B. Manheim and Richard C. Rich, *Empirical Political Analysis: Research Methods in Political Science* (Englewood Cliffs, N.J.: Prentice-Hall, 1981), pp. 59–66.

2. See Andrew S. McFarland, *Common Cause: Lobbying in the Public Interest* (Chatham, N.J.: Chatham House, 1984).

3. See Kay Lehman Schlozman and John T. Tierney, *Organized Interests and American Democracy* (New York: Harper and Row, 1986).

4. Interview John Strohm, executive editor, *National Wildlife* magazine, October 23, 1985, Washington, D.C., National Wildlife Federation Washington Headquarters.

5. Interview John Gottschalk, member of the board of directors, National Wildlife Federation, October 23, 1985, Washington, D.C., International Association of Fish and Wildlife Agencies Headquarters.

6. Interview, Jim Keough, editor, *Sierra* magazine, December 19, 1985, San Francisco, California, Sierra Club National Headquarters.

7. Interview, Patricia Byrnes, managing editor, *Wilderness* magazine, October 17, 1985, Washington, D.C., The Wilderness Society National Headquarters.

8. Interview, Rose Audette, editor, *Environmental Action* magazine, October 21, 1985, Washington, D.C., Environmental Action National Headquarters.

"1998 Earth Share Member Organizations." *Earth Share* (Washington, D.C., Fall 1998), pp. 1–7.

AIP Charity Rating Guide and Watchdog Report (Summer/Fall 1994).

AIP Charity Rating Guide and Watchdog Report (March 1996).

Aldrich, Howard E. *Organizations and Environments.* Englewood Cliffs, N.J.: Prentice-Hall, 1979.

Alexander, Jeffrey C. "Formal and Substantive Voluntarism in the Work of Talcott Parsons: A Theoretical and Ideological Reinterpretation." *American Sociological Review* 43 (1978): 177–198.

Allen, Scott. "The Greening of the Movement: Big Money is Bankrolling Select Environmental Causes." *Boston Globe,* October 19, 1997, p. A1.

Anderson, Sarah, Paul Brotherton, and Peter L. Kelley, eds. *The Scorecard.* Washington, D.C.: League of Conservation Voters, 1994.

Argyris, Chris et al., eds. *Regulating Business.* San Francisco: Institute for Contemporary Studies, 1978.

Asbell, Bernard. *The Senate Nobody Knows.* Garden City, N.Y.: Doubleday, 1978).

Asher, Herbert B. et al., eds. *Theory Building and Data Analysis in the Social Sciences.* Knoxville: University of Tennessee Press, 1984.

"Authors of 'Green Scissors' Report Findings at Briefing." *BNA National Environment Daily,* March 20, 1995, pp. 1–2.

Bacharach, Samuel B., ed. *Research in the Sociology of Organizations.* Greenwich, Conn.: JAI Press, 1981.

Baker, Matthew Reed. "What Interest Groups Pay Their Leaders." *National Journal,* April 26, 1997, pp. 814–819.

Baldwin, Deborah. "Ideology by Mail." *New Republic,* July 7–14, 1979, p. 20.

Barnum, Alex. "A Fresh Look for Sierra Club." *San Francisco Chronicle,* May 25, 1996, p. A1.

Bauer, Raymond A., Ithiel de Sola Pool, and Lewis A. Dexter. *American Business and Public Policy.* New York: Atherton Press, 1963.

Bendavid, Naftali. "Environmental Inaction: Death of an Environmental Group."
    Legal Times, December 2, 1996, p. 1.
Bergman, B. J. "Majority Rules, And It's Green." *Sierra,* January/February 1997,
    pp. 50, 52–53.
——. "Reaching Out, One by One." *Sierra,* January/February 1997,
    p. 53.
Berry, Jeffrey M. *The Interest Group Society,* 2nd ed. Boston: Scott,
    Foresman/Little, Brown, 1989.
——. *The Interest Group Society.* Boston: Little, Brown, 1984.
——. *Lobbying for the People: The Political Behavior of Interest Groups.* Princeton:
    Princeton University Press, 1977.
——. "On the Origins of Public Interest Groups: A Test of Two Theories." *Polity*
    10 (1978): 379–397.
Berry, Jeffrey M., Kent E. Portney, and Ken Thomson. *The Rebirth of Urban
    Democracy.* Washington, D.C.: Brookings Institution, 1993.
Bisbort, Alan. "Can't See the Forest for the Junk Mail." *Washington Post,* March 7,
    1993, p. C5.
Blau, Peter M. *Structure of Organizations.* New York: Basic Books, 1971.
Boerner, Christopher and Jennifer Chilton Kallery. "Restructuring Environmental
    Big Business." St. Louis: Center for the Study of American Business,
    Washington University, December, 1994.
Bonner and Associates/Gallup Poll. "Survey of Business Issues in the New
    Congress." January 1993, pp. 10–11.
Bookchin, Murray. *The Rise of Urbanization and the Decline of Citizenship.* San
    Francisco: Sierra Club Books, 1987.
Borochoff, Daniel. "The Latest Patterns of Telemarketing Abuse." *AIP Charity
    Rating Guide and Watchdog Report* (Winter 1995), p. 1.
——. "Overview: How Americans Give." *AIP Charity Rating Guide and Watchdog
    Report* (Summer/Fall 1994), p. 15.
——. "Think! . . . Before You Give." *AIP Charity Rating Guide and Watchdog
    Report* (Summer/Fall 1994), pp. 1–2.
Borrelli, Peter. "Environmentalism at a Crossroads." In Peter Borrelli, ed.,
    Crossroads: Environmental Priorities for the Future, pp. 3–26. Washington,
    D.C.: Island Press, 1988.
Bosso, Christopher J. "Seizing Back the Day: The Challenge of Environmental
    Activism in the 1990s." In Norman J. Vig and Michael E. Kraft, eds.,
    Environmental Policy in the 1990s, pp. 53–74. 3rd ed. Washington, D.C.:
    Congressional Quarterly Press, 1997.
——. "The Color of Money: Environmental Groups and the Pathologies of Fund
    Raising." In Allan J. Cigler and Burdett A. Loomis, eds., *Interest Group
    Politics,* pp. 101–130. 4th ed. Washington, D.C.: Congressional Quarterly
    Press, 1995.
——. "After the Movement: Environmental Activism in the 1990s." In Norman J.

Vig and Michael E. Kraft, eds., *Environmental Policy in the 1990s,* pp. 31–50. Washington, D.C.: Congressional Quarterly Press, 1994.

——. "Adaptation and Change in the Environmental Movement." In Allan J. Cigler and Burdett A. Loomis, eds., *Interest Group Politics,* pp. 151–176. 3rd ed. Washington, D.C.: Congressional Quarterly Press, 1991.

——. *Pesticides and Policies: The Life Cycle of a Public Issue.* Pittsburgh: University of Pittsburgh Press, 1987.

Bozeman, Barry. *All Organizations Are Public: Bridging Public and Private Organizational Theories.* San Francisco: Jossey-Bass, 1987.

Branigin, William. "Sierra Club Votes for Neutrality on Immigration: Population Issue 'Intensely Debated.'" *Washington Post,* April 26, 1998, p. A16.

Brown, Clifford W., Jr., Lynda W. Powell, and Clyde Wilcox. *Serious Money: Fundraising and Contributing in Presidential Nominating Campaigns.* New York: Cambridge University Press, 1995.

Browne, William P. "Organized Interests, Grassroots Confidants, and Congress." In Allan J. Cigler and Burdett A. Loomis, eds., *Interest Group Politics,* pp. 281–298. 4th ed. Washington, D.C.: Congressional Quarterly Press.

——. "Benefits and Membership: A Reappraisal of Interest Group Activity." *Western Political Quarterly* 29 (1976): 58–273.

Browning, Graeme. *Electronic Democracy Using the Internet in American Politics.* Wilton, Conn.: Online, Pemberton Press, 1996.

Burke, Edmund. "On Election to Parliament." Reprinted in Diane Ravitch and Abigail Thernstrom, eds., *The Democracy Reader.* New York: HarperCollins, 1992.

Burns, James MacGregor. *Leadership.* New York: Harper and Row, 1978.

Burson-Marsteller. "Communications and Congress: A Study of the Exposure of Federal Legislative Offices to Various Information Vehicles." Washington, D.C.: Public Affairs/Government Relations, Burson-Marsteller, 1981.

Butterfield, Stephen. *Amway: The Cult of Free Enterprise.* Boston: South End Press, 1985.

Bykerk, Loree and Ardith Maney. "Consumer Groups and Coalition Politics on Capitol Hill." In Allan J. Cigler and Burdett A. Loomis, eds., *Interest Group Politics,* pp. 259–280. 4th ed. Washington, D.C.: Congressional Quarterly Press, 1983.

Cahn, Robert, ed. *An Environmental Agenda for the Future.* Washington, D.C.: Island Press, 1985.

Camisa, Catalina. "Poor Minorities Want Voice in Environmental Choices." *Congressional Quarterly Weekly Report,* August 21, 1993, pp. 2257–2260.

Campbell, Colton and Roger Davidson. "Coalition Building in Congress." In Paul S. Herrnson, Ronald G. Shaiko, and Clyde Wilcox, eds., *The Interest Group Connection: Electioneering, Lobbying, and Policymaking in Washington,* pp. 116–136. Chatham, N.J.: Chatham House, 1998.

Cannizaro, Steve. "Con Men Target Elderly in Roof Repair Scam." *New Orleans Times-Picayune,* January 25, 1996, p. B1.

Carr, Patrick, ed. *Sierra Club: A Guide.* San Francisco: Sierra Books, 1989.

Carson, Rachel. *Silent Spring.* Boston: Houghton Mifflin, 1962.

Chamberlain, John R. "Provision of Collective Goods as a Function of Group Size." *American Political Science Review* 68 (1974): 707–716.

Chappie, Damon. "The IRS's 'Story of M' May Affect '96 Politics." *Roll Call,* April 15, 1996, pp. 1, 31.

Chetwynd, Josh. "Splatter-Casting the Sierra Club." *U.S. News and World Report,* March 31, 1997, p. 37.

Cigler, Allan J. and Burdett A. Loomis. "Contemporary Interest Group Politics: More than 'More of the Same.'" In Allan J. Cigler and Burdett A. Loomis, eds., *Interest Group Politics,* pp. 393–406. 4th ed. Washington, D.C.: Congressional Quarterly Press, 1995.

Cigler, Allan J. and Burdett A. Loomis, eds. *Interest Group Politics.* Washington, D.C.: Congressional Quarterly Press, 1983.

——, eds. *Interest Group Politics.* 2nd ed. Washington, D.C.: Congressional Quarterly Press, 1986.

——, eds. *Interest Group Politics.* 3rd ed. Washington, D.C.: Congressional Quarterly Press, 1991.

——, eds. *Interest Group Politics.* 4th ed. Washington, D.C.: Congressional Quarterly Press, 1995.

Cigler, Allan J. and Anthony J. Nownes. "Public Interest Entrepreneurs and Group Patrons." In Allan J. Cigler and Burdett A. Loomis, eds., *Interest Group Politics,* pp. 77–100. 4th ed. Washington, D.C.: Congressional Quarterly Press, 1995.

Clapp, Charles E. *The Congressman: His Work as He Sees It.* Washington, D.C.: Brookings Institution, 1963.

Clark, Peter B. and James Q. Wilson. "Incentive Systems: A Theory of Organizations." *Administrative Science Quarterly* 6 (1961): 29–166.

Clark, Timothy B. "After a Decade of Doing Battle, Public Interest Groups Show Their Age." *National Journal,* July 12, 1980, pp. 1136–1141.

Clifford, Frank. "Environmental Movement Struggles as Clout Fades." *Los Angeles Times,* September 21, 1994, p. A1.

Clotfelter, Charles T. "Charitable Giving and Tax Legislation in the Reagan Era." *Law and Contemporary Problems* 48 (1985): 197–212.

Cockburn, Alexander. "The Green Betrayers: Major Environmental Groups and the Clinton Administration." *The Nation,* February 6, 1995, p. 157.

——. "Wilderness Society: New Shame." *The Nation,* February 20, 1995, p. 228.

——. "Wilderness Society: The Saga of Shame Continues." *The Nation,* March 6, 1995, p. 300.

——. "Wilderness Society: Saga of Shame (Part MMDVI)." *The Nation,* April 3, 1995, p. 444.

———. "Roush: The Sequel." *The Nation,* May 1, 1995, pp. 588–589.

Cockburn, Alexander and Jeffrey St. Clair. "Wilderness Chief in Tree Massacre." *The Nation,* April 24, 1995, p. 556–589.

Cohen, Bernard. *The Press and Foreign Policy.* Princeton: Princeton University Press, 1963.

Cohen, David. "The 1990s and the Public Interest Movement." In Leslie Swift-Rosenzweig, ed. *Public Interest Profiles, 1988–1989,* pp. xi–xvii. Washington, D.C.: Congressional Quarterly Press, 1988.

———. "Future Directions for the Public Interest Movement." *Citizen Participation* (March/April 1982): 3–4, 22.

Cohen, Jean L. "Strategy or Identity: New Theoretical Paradigms and Contemporary Social Movements." *Social Research* 52 (1985): 63–716.

Cohen, Michael P. "Origins and Early Outings." In Patrick Carr, ed. *The Sierra Club: A Guide,* pp. 9–17. San Francisco: Sierra Club Books, 1989.

Cohen, Richard E. *Washington at Work: Back Rooms and Clean Air.* New York: Macmillan, 1992.

Collett, Jonathan and Stephen Karakashian. *Greening the College Curriculum: A Guide to Environmental Teaching in the Liberal Arts.* Washington, D.C.: Island Press, 1996.

Commoner, Barry. *Science and Survival.* New York: Ballantine Books, 1963.

Conca, Ken, Michael Alberty, and Geoffrey D. Dabelko, eds. *Green Planet Blues: Environmental Politics from Stockholm to Rio.* Boulder: Westview Press, 1995.

Conway, M. Margaret. *Political Participation in the United States.* Washington, D.C.: Congressional Quarterly Press, 1985.

Cook, Constance Ewing. "Participation in Public Interest Groups: Membership Motivations." *American Politics Quarterly* 12 (1984): 409–430.

Cosmedy, Alison, ed. *DMA 1994/1995 Statistical Fact Book.* Washington, D.C.: Direct Marketing Association, 1994.

———, ed. *DMA 1995/1996 Statistical Fact Book.* Washington, D.C.: Direct Marketing Association, 1995.

Costain, Anne N. "Representing Women: The Transition from Social Movement to Interest Group." In Ellen Boneparth, ed., *Women, Power, and Policy,* pp. 19–37. Elmsford, N.Y.: Pergamon Press, 1982.

Costain, Douglas W. and Anne N. Costain. "Representing Women: The Transition from Social Movement to Interest Group." *Western Political Quarterly* 34 (1981): 100–113.

Crandall, Robert. "The Use of Environmental Policy to Reduce Economic Growth in the Sunbelt: The Role of Electric Utility Rates." In Michael Crew, ed., *Regulatory Reform and Public Utilities,* pp. 125–140. Lexington, Mass.: Lexington Books, 1982.

Crew, Michael, ed. *Regulatory Reform and Public Utilities.* Lexington, Mass.: Lexington Books, 1982.

Cunningham, Noble E. *Circular Letters of Congressmen to Their Constituents, 1789–1879*. Chapel Hill: University of North Carolina Press, 1978.

Dahl, Robert A. *Dilemmas of Pluralist Democracy: Autonomy versus Control*. New Haven: Yale University Press, 1982.

DeBell, Garrett, ed. *The New Environmental Handbook*. San Francisco: Friends of the Earth, 1980.

Dembner, Alice. "Movement Is Strong on Campus." *Boston Globe*, November 12, 1994, p. 28.

Devall, William Bert. "The Governing of a Voluntary Organization: Oligarchy and Democracy in the Sierra Club." Ph.D. diss., University of Oregon, 1970.

Devroy, Ann. "Environmental Expert Steals Show at Deregulation Party." *Washington Post*, April 30, 1992, p. A24.

"Dialogue II: Papers Call Deal 'Unusual'; Enviros Cautious." *Greenwire*, July 9, 1992, p. 1.

"The DISCRIM Procedure." In Alice Allen Ray, ed., *SAS User's Guide*, pp. 381–396. Cary, N.C.: SAS Institute, 1982.

Dowie, Mark. *Losing Ground: American Environmentalism at the Close of the Twentieth Century*. Cambridge, Mass.: MIT Press, 1995.

———. "The Fourth Wave: Environmental Movement's Evolution." *Mother Jones*, March 1995, pp. 34–38.

Downs, Anthony. "Up and Down with Ecology: The Issue Attention Cycle." *Public Interest* 94 (Summer 1972): 28–38.

Dudley, Barbara. "It's Not Easy Being Green." *Mother Jones*, May/June 1995, p. 8.

Duerst-Lahti, Georgia. "The Government's Role in Building the Women's Movement." *Political Science Quarterly* 104 (1989): 249–268.

Dunlap, Riley E. and Michael Allen. "Partisan Differences on Environmental Issues: A Congressional Roll Call Analysis." *Western Political Quarterly* 29 (1976): 384–397.

Dunlap, Riley E. and Angela G. Mertig. "The Evolution of U.S. Environmental Movement from 1970 to 1990: An Overview." In Riley E. Dunlap and Angela G. Mertig, eds., *American Environmentalism: The U.S. Environmental Movement, 1970–1990*, pp. 1–10. Philadelphia: Taylor and Francis, 1992.

"EDF, AD Council, McDonald's Team Up to Encourage Purchase of Recycled Products." *BNA National Environment Daily*, January 12, 1995, p. 1–5.

"EDF Challenges 'Win-Lose' Thinking." *BNA Environmental Law Update*, July 20, 1992, p. 3.

Ehrlich, Paul R. *The Population Bomb*. New York: Ballantine Books, 1968.

Eisenberg, Pablo. "Community/Grassroots Organizations." In Douglas J. Bergner, ed., *Public Interest Profiles*, pp. 19–26. Washington, D.C.: Foundation for Public Affairs, 1986.

Elder, Charles D. and Roger W. Cobb. *The Political Uses of Symbols.* New York: Longman, 1983.

Engelberg, Stephen. "A New Breed of Hired Hands Cultivates Grass-Roots Anger." *New York Times,* March 17, 1993, pp. A1, A17.

Enda, Jodi. "Organizing the Grass Roots to Plant a Seed in Washington." *Philadelphia Inquirer,* October 15, 1995, pp. A1, A8.

Ensman, Richard G., Jr. "The Art and Science of Direct Mail Copywriting: Part One." *Grassroots Fundraising Journal* 6 (April 1987): 3–14.

"Enviro Groups: Washington University Study Stirs Debate in Community." *Greenwire,* January 4, 1995, p. 1.

Environmental Action. "1985 Annual Report: Environmental Action/ Environmental Action Foundation." Washington, D.C.: Environmental Action, 1985.

"Environmental Groups Launch Campaign to Counter Threats to Pending Legislation." *BNA National Environment Daily,* July 8, 1994, pp. 1–4.

"Environmental Organizations Join Coalition to Oppose Federal Subsidies." *BNA National Environmental Daily,* March 25, 1993, pp. 1–2.

Etzioni, Amitai. *A Comparative Analysis of Complex Organizations.* New York: Free Press, 1975.

Fenno, Richard F., Jr. *Homestyle: House Members in their Districts.* Boston: Little, Brown, 1978.

Ferguson, Kathleen. "Toward a Geography of Environmentalism in the United States." M.A. thesis, California State University, Hayward, 1985.

Fiorina, Morris P. "Formal Models in Political Science." *American Journal of Political Science* 19 (1975): 133–159.

Forsythe, David P. and Susan Welch. "Joining and Supporting Public Interest Groups: A Note on Some Empirical Findings." *Western Political Quarterly* 36 (1983): 386–399.

Foundation for Public Affairs. *Public Interest Profiles: 1982–1983.* Ed. John F. Mancini. 3rd ed. Washington, D.C.: Foundation for Public Affairs, 1982.

——. *Public Interest Profiles: 1984–1985.* Ed. John Mancini. 4th ed. Washington, D.C.: Foundation for Public Affairs, 1984.

——. *Public Interest Profiles: 1986–1987.* Ed. Douglas J. Bergner. 5th ed. Washington, D.C.: Foundation for Public Affairs, 1986.

——. *Public Interest Profiles: 1988–1989.* Ed. Leslie Swift-Rosenzweig. 6th ed. Washington, D.C.: Congressional Quarterly Press, 1988.

——. *Public Interest Profiles: 1992–1993.* Ed. Leslie Swift-Rosenzweig. 7th ed. Washington, D.C.: Congressional Quarterly Press, 1992.

——. *Public Interest Profiles: 1996–1997.* Ed. Paul McClure. 8th ed. Washington, D.C.: Congressional Quarterly Press, 1996.

Fowler, Linda L. and Ronald G. Shaiko. "The Grass Roots Connection: Environmental Activists and Senate Roll Calls." *American Journal of Political Science* 31 (1987): 484–510.

Fox, Stephen. *The American Conservation Movement*. Madison: University of
    Wisconsin Press, 1985.
——. "We Want No Straddlers." *Wilderness* 48 (1985): 5–19.
——. *John Muir and His Legacy*. Boston: Little, Brown, 1981.
Freeman, Jo. *The Politics of Women's Liberation*. New York: David McKay, 1975.
Fried, John J. "Major Environmental Groups Losing Members, Money." New
    Orleans *The Times-Picayune,* December 25, 1994, p. A3.
Frolich, Norman, Joe Oppenheimer, and Oran Young. *Political Leadership and
    Collective Goods*. Princeton: Princeton University Press, 1971.
Gale, Richard P. "Social Movements and the State." *Sociological Perspectives* 29
    (1986): 202–240.
Gamson, William A. *The Strategy of Social Protest*. Homewood, Ill.: Dorsey Press,
    1975.
Gardner, John W. *On Leadership*. New York: Free Press, 1990.
"Getting the Most from Your Rating Guide." *AIP Charity Rating Guide and
    Watchdog Report* (Summer/Fall 1994), pp. 3–4.
Gifford, Bill. "Merchandise the Planet: The National Wildlife Federation Profits
    by Nature." *Washington City Paper,* April 24, 1992, pp. 10, 12.
Gimpel, James G. "Grassroots Organizations and Equilibrium Cycles in Group
    Mobilization and Access." In Paul S. Herrnson, Ronald G. Shaiko, and
    Clyde Wilcox, eds., *The Interest Group Connection: Electioneering,
    Lobbying, and Policymaking in Washington,* pp. 100–115. Chatham, N.J.:
    Chatham House, 1998.
Ginsburg, Marsha. "Sierra Club to Slash Its Staff." *San Francisco Examiner,*
    November 18, 1994, p. B4.
Godwin, R. Kenneth. "Lobbying Choices of Citizen Action Groups."
    Manuscript, 1984.
——. *One Billion Dollars of Influence: The Direct Marketing of Politics*. Chatham,
    N.J.: Chatham House, 1988.
Godwin, R. Kenneth and Robert C. Mitchell. "Rational Models, Collective
    Goods, and Nonelectoral Political Behavior." *Western Political Quarterly* 35
    (1982): 161–180.
——. "The Impact of Direct Mail on Political Organizations." *Social Science
    Quarterly* 65 (1984): 829–839.
Gordon, C. Wayne and Nicholas Babchuk. "A Typology of Voluntary
    Associations." *American Sociological Review* 24 (1959): 22–29.
Gordon, Rue E., ed. *1993 Conservation Directory*. Washington, D.C.: National
    Wildlife Federation, 1993.
Gottleib, Robert. *Forcing the Spring: The Transformation of the American
    Environmental Movement*. Washington, D.C.: Island Press, 1993.
Gurr, Ted Robert. *Why Men Rebel*. Princeton: Princeton University Press, 1970.
Hair, Jay D. "The Earth's Environment: A Legacy in Jeopardy." In Peter Borrelli,
    ed., *Crossroads: Environmental Priorities for the Future,* pp. 199–206.
    Washington, D.C.: Island Press, 1988.

Hair, Jay D. and Patrick A. Parenteau. "Special Report: The Results of the National Wildlife Federation's Associate Membership Survey and Affiliate Leadership Poll on Major Conservation Policies." Washington, D.C.: National Wildlife Federation, July 14, 1981.

Hansen, John Mark. *Gaining Access: Congress and the Farm Lobby, 1919–1981.* Chicago: University of Chicago Press, 1991.

——. "The Political Economy of Group Membership." *American Political Science Review* 79 (1985): 79–96.

Hansmann, Henry B. "The Role of Nonprofit Enterprise." *The Yale Law Journal* 89 (1980): 835–851.

Harbison, John R. and Peter Pekar Jr. *A Practical Guide for Leapfrogging the Learning Curve.* Los Angeles: Booz-Allen and Hamilton, 1993.

Hardin, Garrett. "The Tragedy of the Commons." *Science* 162 (1968): 1243–1255.

Hardin, Russell. *Collective Action.* Baltimore: Johns Hopkins University Press, 1982.

——. "Collective Action as an Agreeable n-Prisoner's Dilemma." *Behavioral Science* 16 (1971): 472–481.

Harrell, Frank E., Jr. "The LOGIST Procedure." In *SUGI Supplemental Library User's Guide,* pp. 181–202. Cary, N.C.: SAS Institute, 1983.

Harris, Jean. "High-Power Words." *Direct Marketing* 51 (1989): 51, 79–80.

Hayes, Michael T. "Interest Groups: Pluralism or Mass Society." In Allan J. Cigler and Burdett A. Loomis, eds., *Interest Group Politics,* pp. 110–125. Washington, D.C.: Congressional Quarterly Press, 1983.

Hays, Samuel P. "Three Decades of Environmental Politics: The Historical Context." In Michael J. Lacey, ed., *Government and Environmental Politics: Essays on Historical Developments Since World War II,* pp. 19–80. Baltimore: Johns Hopkins University Press, 1991.

——. *Beauty, Health, and Permanence: Environmental Politics in the United States, 1955–1985.* Cambridge, Mass.: Cambridge University Press, 1987.

Herring, Pendleton. *Group Representation Before Congress.* Baltimore: Johns Hopkins University Press, 1929.

Herrnson, Paul S., Ronald G. Shaiko, and Clyde Wilcox, eds. *The Interest Group Connection: Electioneering, Lobbying, and Policymaking in Washington.* Chatham, N.J.: Chatham House, 1998.

Himelfarb, Richard. *Catastrophic Politics: The Rise and Fall of the Medicare Catastrophic Coverage Act of 1988.* University Park: Pennsylvania State University Press, 1995.

Hirschman, Albert O. *Exit, Voice, and Loyalty.* Cambridge, Mass.: Harvard University Press, 1970.

——. *Shifting Involvements: Private Interests and Public Action.* Princeton: Princeton University Press, 1982.

Hochschild, Arlie Russell. "Emotion Work, Feeling Rules, and Social Structure." *American Journal of Sociology* 85 (1979): 551–574.

Hofstadter, Richard. *The Age of Reform.* New York: Vintage Books, 1955.

Hopkins, Bruce R. *The Law of Tax Exempt Organizations*. 2nd ed. Washington, D.C.: Lerner, 1977.

Hrebenar, Ronald J. and Ruth K. Scott. *Interest Group Politics in America*. Englewood Cliffs, N.J.: Prentice Hall, 1990.

Huelskamp, Timothy A. "Congressional Change: Committees on Agriculture in the U.S. Congress." Ph.D. diss., American University, 1995.

Hula, Kevin. "Rounding Up the Usual Suspects: Forging Interest Group Coalitions in Washington." In Allan J. Cigler and Burdett A. Loomis, eds., *Interest Group Politics*, pp. 239–258. 4th ed. Washington, D.C.: Congressional Quarterly Press, 1995.

Imig, Douglas. "Resource Mobilization and Survival Tactics of Poverty Advocacy Groups." *Western Political Quarterly* 45 (June 1992): 501–520.

Internal Revenue Service. "Tax-Exempt Status For Your Organization." Publication 557. Washington, D.C.: Department of the Treasury, Rev. July 1985.

Isaak, Alan C. *Scope and Methods of Political Science*. Chicago: Dorsey Press, 1984.

"Issues Update: Nonprofit Mail." *Association Management* 50 (1) (January 1998): 9.

"Istook-McIntosh-Ehrlich Proposal: Hearing before the Subcommittee on National Economic Growth, Natural Resources, and Regulatory Affairs of the Committee on Government Reform and Oversight." U.S. House of Representatives. 104th Cong., 1st sess., September 28, 1995.

Iyengar, Shanto and Donald Kinder. *News that Matters: Television and American Opinion*. Chicago: University of Chicago Press, 1987.

Jackson, John E. *Constituents and Leaders in Congress*. Cambridge, Mass.: Harvard University Press, 1974.

——. "Measuring the Demand for Environmental Quality Using Survey Data." *The Journal of Politics* 45 (1983): 335–350.

Johnson, Janet B. and Richard A. Joslyn. *Political Science Research Methods*. Washington, D.C.: Congressional Quarterly, 1986.

Jolson, Marvin A. *Consumer Attitudes Toward Direct-to-Home Marketing Systems*. New York: Dunellen, 1970.

Jones, Charles O. *Clean Air: The Policies and Politics of Pollution Control*. Pittsburgh: University of Pittsburgh Press, 1975.

Kau, James B. and Paul H. Rubin. "Public Interest Lobbies: Membership and Influence." *Public Choice* 34 (1979): 45–54.

Keller, Bill. "Computers and Laser Printers Have Recast the Injunction: 'Write Your Congressman.'" *Congressional Quarterly Weekly Report*, September 12, 1982, p. 2247.

——. "Environmental Movement Checks Its Pulse and Finds Obituaries Are Premature." *Congressional Quarterly Weekly Report*, January 31, 1981, pp. 211–216.

——. "Lowest Common Denominator Lobbying: Why the Banks Fought Withholding." *Washington Monthly*, May 1983, pp. 32–29.

———. "Special-Interest Lobbies Cultivate the 'Grassroots' to Influence Capitol Hill." *Congressional Quarterly Weekly Report,* September 12, 1981, p. 1741.

Keniry, Julian. "Environmental Movement Booming on Campus." *Change,* September/October 1993, p. 42.

Kenworthy, Tom. "Conservationist's Logging Deal Draws Fire." *Los Angeles Times,* April 7, 1995, p. A41.

———. "Wilderness Society President Sold Timber Cut on His Montana Ranch." *Washington Post,* April 7, 1995, p. A28.

Keohane, Robert O. and Joseph S. Nye Jr. "Power and Interdependence in the Information Age." *Foreign Affairs* 77 (5) (September/October 1998): 81–94.

Key, Valdimer O., Jr. *The Responsible Electorate: Rationality and Presidential Voting, 1936–1960.* New York: Vintage Books, 1966.

King, David C. and Jack L. Walker. "The Provision of Benefits by Interest Groups in the United States." *Journal of Politics* 54 (1992): 394–426.

Kingdon, John. *Agendas, Alternatives, and Public Policies.* Boston: Little, Brown, 1984.

———. *Congressmen's Voting Decisions.* New York: Harper and Row, 1973.

Kinsley, Michael. "The Moral Myopia of Magazines." *Washington Monthly* 7 (September 1975): 7–18.

Kirk, W. Astor. *Nonprofit Organization Governance.* New York: Carlton Press, 1986.

Klein, Kim. "Twenty Words that Sell." *Grassroots Fundraising Journal* 6 (1987): 13–14.

Kleiner, Art. "The Greening of Jay Hair." *Garbage* (January/February 1991): 3.

Knickerbocker, Brad. "Environment vs. Immigrants." *The Christian Science Monitor,* April 27, 1998, p. 3.

———. "Nation's Green Advocates See Their Groups on Critical List." *Christian Science Monitor,* October 17, 1994, p. 1.

Knoke, David. *Organizing for Collective Action: The Political Economies of Associations.* New York: Aldine de Gruyter, 1990.

———. "Commitment and Detachment in Voluntary Associations." *American Sociological Review* 46 (1981): 141–158.

Knoke, David and David Prensky. "What Relevance Do Organization Theories Have for Voluntary Associations?" *Social Science Quarterly* 65 (1984): 3–20.

Knoke, David and James R. Wood. *Organized for Action: Commitment in Voluntary Associations.* New Brunswick, N.J.: Rutgers University Press, 1981.

Knoke, David and Christine Wright-Isak. "Individual Motives and Organizational Incentive Systems." In Samuel B. Bacharach, ed., *Research in the Sociology of Organizations,* pp. 209–254. Greenwich, Conn.: JAI Press, 1982.

Koehl, Carla and Marc Peyser. "Green Revolt." *Newsweek,* July 10, 1995, p. 6.

Kornhauser, William. *The Politics of Mass Society.* New York: Free Press, 1959.

Kriesberg, Louis, ed. *Research in Social Movements, Conflict, and Change.*
Greenwich, Conn.: JAI Press, 1981.
——, ed. *Research in Social Movements, Conflict, and Change.* Greenwich, Conn.:
JAI Press, 1979.
Kriz, Margaret. "Getting Ready for Round Two." *National Journal,* March 11,
1995, p. 644.
——. "Mark Van Putten: Wildlife Lobby Returns to Its Roots." *National Journal,*
January 25, 1997, p. 184.
——. "The Conquered Coalition." *National Journal,* December 14, 1994, p. 2826.
Krupp, Fred. "Director's Message: Please, Mr. Postman." *EDF Letter,* March 1996,
p. 3.
Kurtz, Howard. "Audubon Chief Pulls Article on Clinton." *Washington Post,*
February 10, 1996, p. C7.
Kuttner, Robert. "Wired for Democracy." *Boston Globe,* February 10, 1996, p. 15.
Ladd, Everett Carll and Karlyn H. Bowman. *Attitudes Toward the Environment:
Twenty-Five Years After Earth Day.* Washington, D.C.: AEI Press, 1995.
Lee, Gary. "Environmentalists Try to Regroup." *Washington Post,* April 22, 1995,
p. A3.
Lee, Martha F. *Earth First! Environmental Apocalypse.* Syracuse, N.Y.: Syracuse
University Press, 1995.
Levin, Meredith, ed. *Lobbying Techniques for the '90s: Strategies, Coalitions, and
Grass-Roots Campaigns.* Washington, D.C.: Congressional Quarterly,
1991.
Lindblom, Charles E. "The Science of Muddling Through." *Public Administration
Review* 19 (1959): 79–88.
Lindblom, Charles E. and David K. Cohen. *Usable Knowledge: Social Science and
Social Problem Solving.* New Haven: Yale University Press, 1979.
Linsky, Martin. *Impact: How the Press Affects Federal Policymaking.* New York:
Norton, 1986.
Lipset, Seymour Martin. *American Exceptionalism: A Double-Edged Sword.* New
York: Norton, 1996.
Lipsky, Michael. "Protest as a Political Resource." *American Political Science
Review* 62 (1968): 1144–1158.
Locksley, Lilu. "The National Geographic: How to Be a Non-Profit and Get
Rich." *Washington Monthly* 9 (September 1977), pp. 46–48.
Loomis, Burdett A. "A New Era: Groups and the Grass Roots." In Allan J. Cigler
and Burdett A. Loomis, eds., *Interest Group Politics,* pp. 169–190.
Washington, D.C.: Congressional Quarterly Press, 1983.
Lowi, Theodore J. *The End of Liberalism.* 2nd ed. New York: Norton, 1979.
——. *The Politics of Disorder.* New York: Basic Books, 1971.
Luttberg, Norman R. *Public Opinion and Public Policy.* Homewood, Ill.: Dorsey
Press, 1974.
Madison, James. "Federalist No. 51." In Alexander Hamilton, James Madison, and

John Jay, *The Federalist Papers,* p. 322. New York: New American Library, Mentor Books, 1961.

Makinson, Larry. *The Price of Admission: Campaign Spending in the 1994 Elections.* Washington, D.C.: Center for Responsive Politics, 1995.

Manes, Christopher. *Green Rage: Radical Environmentalism and the Unmaking of Civilization.* Boston: Little, Brown, 1990.

Manheim, Jarol B. and Richard C. Rich. *Empirical Political Analysis: Research Methods in Political Science.* Englewood Cliffs, N.J.: Prentice-Hall, 1981.

Mansbridge, Jane J. *Why We Lost the ERA.* Chicago: University of Chicago Press, 1986.

March, James G. and Johan P. Olsen. *Ambiguity and Choice in Organizations.* Bergen, Norway: Universitetsforlaget, 1979.

Marcus, Gregory B. "Electoral Coalitions and Senate Roll Call Behavior: An Ecological Analysis." *American Journal of Political Science* 18 (1974): 595–607.

Mardon, Mark. "Where Are We Anyway?" *Sierra,* May/June 1991, pp. 18–19.

Marsh, David. "On Joining Interest Groups." *British Journal of Political Science* 6 (1976): 257–271.

Marwell, Gerald and Ruth E. Ames. "Experiments on the Provision of Public Goods: I. Resources, Interest, Group Size, and the Free Rider Problem." *American Journal of Sociology* 84 (1979): 1334–1360.

McCann, Michael W. *Taking Reform Seriously: Perspectives on Public Interest Liberalism.* Ithaca: Cornell University Press, 1986.

McCarthy, John D. and Mayer N. Zald. "Resource Mobilization and Social Movements: A Partial Theory." *American Journal of Sociology* 82 (1977): 1212–1241.

——. *The Trend in Social Movements in America: Professionalization and Resource Mobilization.* Morristown, N.J.: General Learning Press, 1973.

McCloskey Michael. "Wilderness at the Crossroads, 1945–1970." *Pacific Historical Review* 41 (1972): 346–361.

McConnell, Grant. "The Conservation Movement: Past and Present." *Western Political Quarterly* 7 (1954): 463–578.

McFarland, Andrew S. *Common Cause: Lobbying in the Public Interest.* Chatham, N.J.: Chatham House, 1984.

——. "Interest Groups and Political Time: Cycles in America." *British Journal of Political Science* 21 (July 1991): 257–284.

——. *Public Interest Lobbies: Decision Making on Energy.* Washington, D.C.: American Enterprise Institute, 1976.

——. "Public Interest Lobbies versus Minority Faction." In Allan J. Cigler and Burdett A. Loomis, eds., *Interest Group Politics,* pp. 324–353. Washington, D.C.: Congressional Quarterly Press, 1983.

——. "Social Movements and Theories of American Politics." In Anne N. Costain and Andrew S. McFarland, eds., *Social Movements and American*

Political Institutions, pp. 7–19. Lanham, Md.: Rowman and Littlefield, 1998.

McGuire, Martin C. "Group Size, Group Homogeneity, and the Aggregate Provision of a Pure Public Good Under Cournot Behavior." *Public Choice* 18 (1974): 107–126.

McLaughlin, Curtis P. *The Management of Nonprofit Organizations*. New York: Wiley, 1986.

Meyer, John W. and Brian Rowan. "Institutionalized Organizations: Formal Structure as Myth and Ceremony." *American Journal of Sociology* 83 (1977): 340–363.

Meyer, Marshall W. and M. Craig Brown. "The Process of Bureaucratization." *American Journal of Sociology* 83 (1977): 364–385.

Michels, Robert. "The Iron Law of Oligarchy." In C. Wright Mills, ed., *Images of Man*, pp. 233–261. New York: George Braziller, 1960.

——. *Political Parties*. New York: Free Press, 1962.

Milbrath, Lester W. *Environmentalists: Vanguard for a New Society*. Albany: State University of New York Press, 1984.

Miller, Ken. "Founder of Many Green Groups Scorns Their Complacency." *Gannett News Service*, April 20, 1995, p. 3.

Milne, George R., Easwar S. Iyer, and Sara Gooding-Williams. "Environmental Organization Alliance Relationships Within and Across Nonprofit, Business, and Government Sectors." *Journal of Public Policy and Marketing* (Fall 1996): 375-403.

Mitchell, Robert C. "From Conservation to Environmental Lobbies: The Development of Modern Environmental Lobbies." In Michael J. Lacey, ed., *Government and Environmental Politics: Essays on Historical Developments since World War II*, pp. 81–114. Baltimore: Johns Hopkins University Press, 1991.

——. "National Environmental Lobbies and the Apparent Illogic of Collective Action." In Clifford Russell, ed., *Collective Decision Making*, pp. 87–121. Baltimore: Johns Hopkins University Press, 1979.

Mitchell, Robert C., Angela G. Mertig, and Riley E. Dunlap. "Twenty Years of Environmental Mobilization: Trends Among National Environmental Organizations." In Riley E. Dunlap and Angela G. Mertig, eds., *American Environmentalism: The U.S. Environmental Movement, 1970–1990*, pp. 11–26. Philadelphia: Taylor and Francis, 1992.

Moe, Terry M. "A Calculus of Group Membership." *American Journal of Political Science* 24 (1980): 593–631.

——. "On the Scientific Status of Rational Models." *American Journal of Political Science* 23 (1979): 215–243.

——. *The Organization of Interests: Incentives and the Internal Dynamics of Political Interest Groups*. Chicago: University of Chicago Press, 1980.

Monsma, Stephen V. *When the Sacred and the Secular Mix: Religious Nonprofit*

Organizations and Public Money. Lanham, Md.: Rowman and Littlefield, 1996.

Morgan-Hubbard, Margaret. "Green Backers." *Environmental Action* (Winter 1996): 22–23.

——. "Money and Environmental Groups: How Clean Is 'Green?'" *Environmental Action* (Winter 1996): 20–21.

——. "Small Is Beautiful." *Environmental Action*, June 22, 1993, p. 16.

Mundo, Philip A. *Interest Groups: Cases and Characteristics*. Chicago: Nelson-Hall, 1992.

——. "The Sierra Club." In Mundo, ed., *Interest Groups: Cases and Characteristics*, pp. 167–201. Chicago: Nelson-Hall, 1992.

Nader, Ralph. "Foreword." In Leslie Swift-Rosenzweig, ed., *Public Interest Profiles, 1992–1993*, pp. xi–xvi. 7th ed. Washington, D.C.: Institution for Public Affairs, 1992.

——. *Unsafe at Any Speed: The Designed-In Dangers of the American Automobile*. New York: Grossman, 1965.

National Wildlife Federation, Document #H1501.

Neff, Gina. "Eco-Efficiency: The Business Link to Sustainable Development." *The Nation*, November 7, 1997, p. 50.

"New EDF Project Will Promote Private-Sector Innovations." *EDF Letter*, March 1996, p. 2.

Nolte, Carl. "Sierra Club Lays Off 20, Puts a Freeze on Travel." *San Francisco Chronicle*, July 13, 1991, p. A15.

——. "Sierra Club Settles on 10% Staff Cut." *San Francisco Chronicle*, November 19, 1994, p. B10.

Nownes, Anthony J. and Grant Neeley. "Public Interest Group Entrepreneurship: Disturbances, Patronage, and Personal Sacrifices." Presented at the Annual Meeting of the Midwest Political Science Association, April 6, 1995, Chicago, Illinois.

"NWF: CEO Resigns Amid Alleged 'Internal Conflicts'," *Greenwire*, July 6, 1995, p. 1.

Oberschall, Anthony. *Social Conflicts and Social Movements*. Englewood Cliffs, N.J.: Prentice-Hall, 1973.

Odegard, Peter H. *Pressure Politics: The Story of the Anti-Saloon League*. New York: Columbia University Press, 1928.

Olson, Mancur. *The Logic of Collective Action: Public Goods and the Theory of Groups*. Cambridge, Mass.: Harvard University Press, 1965.

Olson, Mancur and Richard Zeckhauser. "An Economic Theory of Alliances." *Review of Economics and Statistics* 43 (1966): 266–279.

"Opinions of Americans Age 45 and Over on the Medicare Catastrophic Coverage Act." Washington, D.C.: Hamilton, Frederick, and Schneiders for AARP/AARP Research and Data Resources Department, January 1989.

"Orchestrated Mail Does Influence Staff—Who Says So? Staff." *The Congressional Staff Journal* (November/December 1981): 1–8.

Ornstein, Norman J. and Shirley Elder. *Interest Groups, Lobbying, and Policymaking*. Washington, D.C.: Congressional Quarterly Press, 1978.

Osterland, Andrew W. "War Among the Nonprofits." *Financial World*, September 1, 1994, p. 52.

Overberg, Peter and Linda Kanamine. "Green But Not Growing." *USA Today*, October 19, 1994, pp. 8A.

Paehlke, Robert C. *Environmentalism and the Future of Progressive Politics*. New Haven: Yale University Press, 1989.

Page, Benjamin I. and Robert Y. Shapiro. *The Rational Public: Fifty Years of Trends in Americans' Policy Preferences*. Chicago: University of Chicago Press, 1992.

Parris, Thomas M. "A Plethora of Campus Environmental Initiatives on the Net." *Environment* 40 (7) (September 1998): 3.

Parsons, Talcott. *The Structure of Social Action*. New York: Free Press, 1937.

"Participation in Interest Groups High." *Gallup Report* 191 (1981): 45–57.

Patterson, Thomas E. *Out of Order*. New York: Knopf, 1993.

——. *The Mass Media Election: How Americans Choose Their Presidents*. New York: Praeger, 1980.

Pell, Eve. "Movements: Buying In." *Mother Jones*, April/May 1990, p. 23.

Peltzman, Sam. "Constituent Interest and Congressional Voting." *The Journal of Law & Economics* 27 (April 1984): 181–210.

Perrow, Charles. "Members as Resources in Voluntary Associations." In William Rosengren and Mark Lefton, eds., *Organizations and Clients*, pp. 93–116. Columbus: Merrill, 1970.

Pettinico, George. "The Public Interest Paradox." *Sierra*, November/December 1995, pp. 28, 30–31.

Philanthropic Research, Inc. *Guidestar Directory of American Charities: 1996 Index*. Williamsburg, Va.: Philanthropic Research, Inc., 1996.

Pitkin, Hannah F. *The Concept of Representation*. Berkeley: University of California Press, 1967.

Pope, Carl. "Earth's Future and Us." *San Francisco Chronicle*, April 22, 1993, p. A21.

——. "Letter to the Editor: Environmental Slippage Blamed On Media, Congressional Foes." *San Francisco Examiner*, October 14, 1994, p. A22.

Pound, Edward T., Gary Cohen, and Penny Loeb. "Tax Exempt!" *U.S. News and World Report*, October 2, 1995, pp. 36–39, 42–44, 46, 51.

Porter, Gareth and Janet Welsh Brown. *Global Environmental Politics*. 2nd ed. Boulder: Westview Press, 1995.

Powell, Walter W., ed. *The Nonprofit Sector: A Research Handbook*. New Haven: Yale University Press, 1987.

Power, Thomas M. and Paul Rauber. "The Price of Everything." *Sierra*, November/December 1993, p. 88.

Priscoli, Jerry Delli. "The Enduring Myths of Public Involvement." *Citizen Participation,* March/April 1982, pp. 5–6, 20.

Przeworski, Adam and Henry Teune. *The Logic of Comparative Social Inquiry.* New York: Wiley, 1970.

Public Interest Perspectives: The Next Four Years. Washington, D.C.: Public Citizen, 1977.

Rabe, Barry G. *Beyond Nimby: Hazardous Waste Siting in Canada and the United States.* Washington, D.C.: Brookings Institution, 1994.

Rauber, Paul. "Beyond Greenwash." *Sierra,* July 1994, pp. 47–50.

——. "Greenwash, Inc." *Environmental Action* (Summer 1994): 8–9.

Rauch, Jonathan. *Demosclerosis: The Silent Killer of American Government.* New York: Times Books, 1994.

Ray, Alice Allen, ed. *SAS User's Guide: Statistics.* Cary, N.C.: SAS Institute, 1982.

Ridgeway, James. "Greenwashing Earth Day." *Village Voice,* April 25, 1995, pp. 15–16.

Riley, John. "Tax-Exempt Foundations: What Is Legal?" *National Law Journal* 9 (24) (February 23, 1987): 8.

Roe, David. *Dynamos and Virgins.* New York: Random House, 1984.

Rosenau, James N. *Citizenship Between Elections.* New York: Prentice-Hall, 1973.

Rosenbaum, Walter A. *Environmental Politics and Policy.* Washington, D.C.: Congressional Quarterly Press, 1985.

Rosengren, William and Mark Lefton, eds. *Organizations and Clients.* Columbus: Merrill, 1970.

Rosenstone, Steven J. and John Mark Hansen. *Mobilization, Participation, and Democracy in America.* New York: Macmillan, 1993.

Rothenberg, Lawrence S. *Linking Citizens to Government: Interest Group Politics at Common Cause.* New York: Cambridge University Press, 1992.

——. "Organizational Maintenance and the Retention Decision in Groups." *American Political Science Review* 82 (1988): 1129–1152.

Rouder, Susan. "Mobilization by Mail." *Citizen Participation,* September/October 1980, pp. 3–4, 16–17.

Roush, G. Jon. "Conservation's Hour: Is Leadership Ready? In Donald Snow, ed., *Voices from the Environmental Movement: Perspectives for a New Era,* pp. 21–40. Washington, D.C.: Island Press, 1992.

Ruben, Barbara and David Lapp. "On the Road from Earth Day: Environmental Action Looks Toward Its Roots to Shape the Future." *Environmental Action,* June 22, 1993, p. 14.

Russell, Clifford, ed. *Collective Decision Making.* Baltimore: Johns Hopkins University Press, 1979.

Russell, Dick. "The Monkey Wrenchers." In Peter Borrelli, ed., *Crossroads: Environmental Priorities for the Future,* pp. 27–48. Washington, D.C.: Island Press, 1988.

Sachs, Richard C. "The Lobbying Disclosure Act of 1995: A Brief Description." *CRS Report to Congress 96–29GOV*, January 4, 1996, pp. 1–6.

Sale, Kirkpatrick. *The Green Revolution: The American Environmental Movement, 1962–1992.* New York: Hill and Wang, 1993.

——. "The U.S. Green Movement Today." *The Nation*, July 19, 1993, pp. 92–96.

Salisbury, Robert H. "An Exchange Theory of Interest Groups." *Midwest Journal of Political Science* 13 (1969): 1–32.

Salisbury, Robert H. et al. "Triangles, Networks, and Hollow Cores: The Complex Geometry of Washington Interest Representation." In Mark P. Petracca, ed., *The Politics of Interests: Interest Groups Transformed,* pp. 130–149. Boulder: Westview Press, 1992.

Sandler, Todd. *Collective Action: Theory and Applications.* Ann Arbor: University of Michigan Press, 1992.

Schattschneider, Elmer E. *The Semi-Sovereign People.* New York: Holt, Rinehart, and Winston, 1960.

——. *The Struggle for Party Government.* College Park: University of Maryland, 1948.

Scheffer, Victor B. *The Shaping of Environmentalism in America.* Seattle: University of Washington Press, 1991.

Schlozman, Kay Lehman and John T. Tierney. "More of the Same: Washington Pressure Groups in a Decade of Change." *The Journal of Politics* 45 (1983): 351–377.

——. *Organized Interests and American Democracy.* New York: Harper and Row, 1986.

Schnaiberg, Allan. *The Environment: From Surplus to Scarcity.* New York: Oxford University Press, 1980.

Schneider, Keith. "Big Environment Hits a Recession." *New York Times,* January 1, 1995, p. 3–4.

——. "Pushed and Pulled, Environment Inc. Is on the Defensive." *New York Times,* March 29, 1992, pp. 1, 4.

Schubert, Glendon. *The Public Interest.* Glencoe, Ill.: Free Press, 1960.

Scotch, Richard. *From Goodwill to Civil Rights: Transforming Federal Disability Policy.* Philadelphia: Temple University Press, 1985.

Scott, Douglas. "Conservation and the Sierra Club." In Patrick Carr, ed., *The Sierra Club: A Guide,* pp. 18–37. San Francisco: Sierra Club Books, 1989.

Selznick, Philip. *Leadership in Administration.* New York: Harper and Row, 1957.

Shabecoff, Phillip. *A Fierce Green Fire: The American Environmental Movement.* New York: Hill and Wang, 1993.

Shaiko, Ronald G. "Greenpeace U.S.A.: Something Old, New, Borrowed." *The Annals* 528 (Summer 1993): 88–100.

——. "Female Participation in Public Interest Nonprofit Governance: Yet Another Glass Ceiling?" *Nonprofit and Voluntary Sector Quarterly* 25 (3) (September 1996): 302–320.

——. "Lobby Reform: Curing the Mischiefs of Factions?" In James A. Thurber and Roger H. Davidson, eds., *Remaking Congress: Change and Stability in the 1990s,* pp. 156–173. Washington, D.C.: Congressional Quarterly Press, 1995.

——. "More Bang for the Buck: The New Era of Full-Service Public Interest Organizations." In Allan J. Cigler and Burdett A. Loomis, eds., *Interest Group Politics,* pp. 109–29. 3rd ed. Washington D.C.: Congressional Quarterly Press, 1991,

——. "The Public Interest Dilemma: Organizational Maintenance and Political Representation in the Public Interest Sector." Ph.D. diss., Syracuse University, 1989.

——. "Religion, Politics, and Environmental Concern: A Powerful Mix of Passions." *Social Science Quarterly* 68 (1987): 244–262.

——. "Petitioning Government: Constituent Communications and Congressional Responses." Paper presented at the Annual Meeting of the Southern Political Science Association, Charlotte, N.C., November 5–7, 1987.

——. "Public Interest Law and the Pluralist Heaven." Paper presented at the Annual Meeting of the Northeastern Political Science Association, Philadelphia, Pennsylvania, November 17–19, 1983.

——. "Interests Group Activity at the Grass Roots Level: A United Front?" Paper presented at the Annual Meeting of the Southern Political Science Association, Savannah, Georgia, November 1–3, 1984.

Shanoff, Barry. "Environmental Groups Suffer Fate of Big Business." *World Wastes,* March 1995, p. 20.

Shaw, Donald L. and Maxwell E. McCombs. *The Emergence of American Political Issues: The Agenda Setting Function of the Press.* St. Paul, Minn.: West, 1977.

Shively, W. Phillips. *The Craft of Political Research.* 2d ed. Englewood Cliffs, N.J.: Prentice-Hall, 1974.

Shott, Susan. "Emotion and Social Life: A Symbolic Interactionist Analysis." *American Journal of Sociology* 84 (1979): 1317–1334.

Sickle, Dirck Van. *The Ecological Citizen.* New York: Harper and Row, 1971.

"Sierra Club Financial Report." *Sierra,* September/October 1995, pp. 102–103.

"Sierra Club Puts New Legislative Leaders on Notice." *PR Newswire,* November 19, 1994, p. 3.

"Sierra Club Youthful Prez's Style Draws Praise, Dissent." *Greenwire,* March 25, 1997, p. 1.

"Sierra Silliness: Editorial." *Providence-Journal Bulletin,* April 24, 1997, p. 7B.

Smelser, Neil. *Theory of Collective Behavior.* New York: Free Press, 1963.

Smith, V. Kerry. "A Theoretical Analysis of the 'Green Lobby.'" *American Political Science Review* 79 (1985): 132–147.

Smucker, Bob. *The Nonprofit Lobbying Guide: Advocating Your Cause—and Getting Results.* San Francisco: Jossey-Bass, 1991.

Snow, Donald. *Inside the Environmental Movement: Meeting the Leadership Challenge.* Washington, D.C.: Island Press, 1992.

Stanfield, Rochelle L. "Environmental Lobby's Changing of the Guard Is Part of Movement's Evolution." *National Journal,* June 5, 1985, pp. 1350–1353.

———. "The Green Blueprint." *National Journal,* July 22, 1988, pp. 1735–1737.

Starobin, Paul. "Raging Moderates." *National Journal,* May 10, 1997, pp. 914–918.

Stauber, John C. and Sheldon Rampton. "Green PR: Silencing Spring." *Environmental Action* (Winter 1996): 16–19.

Steinhart, Peter. "What Can We Do About Environmental Racism?" *Audubon* 93 (4) (May 1991): 18–20.

Stone, Peter H. "Learning from Nader," *National Journal,* June 11, 1994, pp. 1342–1344.

———. "Payday!" *National Journal,* December 14, 1994, pp. 2948–2960.

Strickler, Karyn. "Environmental Towers Build Too High to Keep Grass Roots." *The Christian Science Monitor,* April 21, 1995, p. 18.

Strohm, John. "Federation Scores High in Lobbying." *National Wildlife* 15 (July/August 1985): 27.

Suttles, Gerald D. and Mayer N. Zald, eds. *The Challenge of Social Control: Citizenship and Institution Building in Modern Society.* Norwood, N.J.: Ablex, 1985.

Swisher, Kara. "There's No Place Like a Home Page." *Washington Post,* July 1, 1996, pp. A1, A8.

Tarrow, Sidney. *Power in Movement: Social Movement, Collective Action, and Politics.* New York: Cambridge University Press, 1994.

Taylor, Wendy. "Enviro Groups: Success at EDF, TNC, and NPCA No Accident." *Greenwire,* December 16, 1994, pp. 1–5.

"The ABCs of Activism." *Sierra,* January/February 1996, p. 23.

Thompson, Grant P. "The Environmental Movement Goes to Business School." *Environment* 27 (May 1985): 7–11, 30.

Thorndike, Jack. "Trouble at the Sierra Club." *The Progressive* 56 (July 1992): 13.

Tillock, Harriet and Denton E. Morrison. "Group Size and Contributions to Collective Action: A Test of Mancur Olson's Theory on Zero Population Growth, Inc." In Louis Kriesberg, ed., *Research in Social Movements, Conflict, and Change,* pp. 131–158. Greenwich, Conn.: JAI Press, 1979.

Tilly, Charles. *From Mobilization to Revolution.* Reading, Mass.: Addison-Wesley, 1978.

Tocqueville, Alexis de. *Democracy in America.* New York: Mentor Books, 1956.

Tousignant, Marylou. "Wildlife Federation's Neighbors Wary of Sale." *Washington Post,* February 12, 1998, p. V1.

Traugott, Michael W. and Paul J. Lavrakas. *The Voter's Guide to Election Polls.* Chatham, N.J.: Chatham House, 1996.

Treusch, Paul E. and Norman A. Sugarman. *Tax-Exempt Charitable*

Organizations. 2d ed. Philadelphia: American Law Institute–American Bar Association, 1983.

Truman, David B. *The Governmental Process.* 2nd ed. New York: Knopf, 1971.

Turner, Ralph H. "Collective Behavior and Resource Mobilization as Approaches to Social Movements: Issues and Continuities." In Louis Kriesberg, ed., *Research in Social Movements, Conflict, and Change,* pp. 8–21. Greenwich, Conn.: JAI Press, 1981.

Turner, Ralph H. and Lewis M. Killian. *Collective Behavior.* Englewood Cliffs, N.J.: Prentice-Hall, 1957.

United States General Accounting Office. "Tax-Exempt Organizations: Information on Selected Types of Organizations." February 1995, GAO/GGD-95-84BR.

United States Census Bureau, 1980 Census Report.

Verba, Sidney, Kay Lehman Schlozman, and Henry E. Brady. *Voice and Equality: Civic Voluntarism in American Politics.* Cambridge, Mass.: Harvard University Press, 1995.

Verba, Sidney and Norman H. Nie. *Participation in America: Political Democracy and Social Equality.* New York: Harper and Row, 1972.

Viguerie, Richard A. *The New Right: We're Ready to Lead.* Falls Church, Va.: Viguerie, 1980.

Walker, Jack L. *Mobilizing Interest Groups in America.* Ann Arbor: University of Michigan Press, 1991.

——. "The Origins and Maintenance of Interest Groups in America." *American Political Science Review* 77 (June 1983): 390–406.

Walsh, Edward J. and Rex H. Warland. "Social Movement Involvement in the Wake of a Nuclear Accident: Activists and Free Riders in the TMI Area." *American Sociological Review* 48 (1983): 764–780.

Wapner, Paul. *Environmental Activism and World Civic Politics.* Albany: State University of New York Press, 1996.

Ward, Bud and Jan Floyd. "Washington Lobbying Groups: How They Rate." *Environmental Forum,* April 1985, pp. 9–18.

Wark, John and Gary Marx. "Faith, Hope, and Chicanery." *Washington Monthly* 19 (January 1987): 29.

Watkins, T. H. and William A. Turnage. "We Still Want No Straddlers." *Wilderness* 48 (1985): 34–37.

Weber, Max. *Economy and Society: An Outline of Interpretive Sociology,* ed. Guenther Roth and Claus Wittich. Berkeley: University of California Press, 1978.

Weisbrod, Burton A. *The Nonprofit Economy.* Cambridge, Mass.: Harvard University Press, 1988.

Wiebe, Robert H. *The Search for Order, 1877–1920.* New York: Hill and Wang, 1967.

Wilderness Society. "Meadows Named New President of Wilderness Society." Press Release. Washington, D.C., September 24, 1996, pp. 1–2.

Wilson, James Q. *Political Organizations*. New York: Basic Books, 1973.

——. *Political Organizations*. Princeton: Princeton University Press, 1995.

Woliver, Laura R. *From Outrage to Action: The Politics of Grassroots Dissent*. Urbana: University of Illinois Press, 1993.

Wright, John R. *Interest Groups and Congress: Lobbying, Contributions, and Influence*. Boston: Allyn and Bacon, 1996.

Young, Dennis R. "Executive Leadership in Nonprofit Organizations." In Walter W. Powell, ed., *The Nonprofit Sector: A Research Handbook*, pp. 167–179. New Haven: Yale University Press, 1987.

Youth, Howard. "Boom Time for Environmental Groups." *WorldWatch*, November-December 1989, p. 33.

Zald, Mayer N. and John D. McCarthy, eds. *The Dynamics of Social Movements: Resource Mobilization, Social Control, and Tactics*. Lanham, Md.: University Press of America, 1988.

——, eds. *Social Movements in an Organizational Society*. New Brunswick, N.J.: Transaction Books, 1987.

Zald, Mayer N. and Roberta Ash. "Social Movement Organizations: Growth, Decay and Change." *Social Forces* 44 (1966): 327–341.

Zey-Ferrell, Mary. "Criticisms of the Dominant Perspective on Organizations." *Sociological Quarterly* 22 (1981): 181–205.